MIXED-ABILITY GROUPING
THEORY AND PRACTICE

A.V. KELLY

D0269968

Harper & Row, Publishers

London   New York   Hagerstown
San Francisco   Sydney

First published 1978
Second impression 1980
Harper & Row Ltd
28 Tavistock Street
London WC2E 7PN

British Library Cataloguing in Publication Data
Kelly, Albert Victor
    Mixed ability grouping. — 2nd ed.
    1. Ability — grouping in education
    I. Title II. Teaching mixed ability classes
    371.2'52   LB3061

ISBN 0-06-318069-3
        0-06-318070-7 Pbk

Cover designed by Richard Dewing, Millions
Typeset by Preface Ltd, Salisbury
Printed by A. Wheaton & Co Ltd, Exeter

Contents

The story is told that, when Michelangelo went to see the Pope who was about to engage him to build the dome of St. Peter's and paint the Sistine Chapel, he brought with him the following testimonial, 'The bearer of these presents is Michelangelo, the sculptor. His nature is such that he needs to be drawn out by kindness and encouragement, but if love be shown him and he be treated really well he could accomplish things that will make the whole world wonder.'

(*Tom Gannon 1975*)

Acknowledgements

My thanks are still due to those who helped in the preparation of the first edition of this book — colleagues in the School of Education at Goldsmiths' College, Meriel Downey, Pamela Moore, Bill Bennett, Graham Byrne Hill, Wynne Davies and John Ellis, and colleagues from elsewhere, John Handford, Roy Schofield and, especially, Arthur Young, who offered equally valuable comments on the draft of this second edition. To this list must now be added the names of Keith Thompson for his advice on this new text and of Geva Blenkin, who helped in the tedious task of preparing an extended bibliography.

My gratitude must also be expressed to the teachers in many parts of the country who have invited me to join in their discussions both of the general issues of mixed-ability teaching and of their own specific concerns. In particular, I must thank those who contributed to the book of case studies which was published to supplement the first edition of this book. I am grateful to them all for the opportunities they gave me to share in their problems and for keeping at the forefront of my own thinking those day-to-day realities that must always be at the heart of any constructive discussion of education. If this new edition shows any improvement on the old, to them must go a large share of the credit.

Vic Kelly

Foreword

The four or five years that have passed since the writing of the first edition of this book, entitled *Teaching Mixed Ability Classes*, have seen both a rapid extension of and, more recently, something of a reaction against mixed-ability groupings, especially in secondary schools. The former is the continuing effect of that realization of the defects and inadequacies of streaming that has grown steadily over the last two decades; the latter is part of that general loss of confidence by teachers in the validity of their professional judgement which has been created by recent demands for greater teacher accountability, demands that have been sparked off by the publicity that has attended one or two cases in which teachers, in attempting to advance the cause of education, have shown themselves to be human, and thus fallible, by getting things wrong.

For two reasons, then, a new edition of this book is needed. In the first place, it is desirable to try to capitalize on the vast experience that has been accumulated over these years as many more schools and teachers have turned to this form of organization and more and more has been learnt from both their successes and their failures. Secondly, it has now become vital to attempt to offer some kind of justification for the mixing of abilities in classes as well as in schools, both for those outside the profession who have genuine doubts about its worth and viability and for teachers themselves, so that they can achieve the kind of understanding of the thinking behind it that will enable them to be confident of its value and of their own way of implementing it. Without either of these no system of any kind can succeed.

The first thing that this new edition attempts to do, therefore, is to take account of some of the things we have learnt about mixed-ability teaching during the last few years while it has been spreading so rapidly in secondary schools. It is now much easier to identify the major concerns of both teachers and headteachers as they face up to a major reorganization of this kind. It is gratifying to note that most of them are those we identified and devoted chapters to in out first edition — the problems of the less able pupil, public examinations, resources and so on. Others, such as the difficulties of introducing mixed-ability grouping into a school for the first time and the problems encountered by particular curriculum subjects, have emerged from the experience of many schools and teachers and have been documented very clearly and illustrated, sometimes quite graphically, in the several collections of case studies that followed the initial publication of this book. This experience and these studies have also provided us with many more examples of the ways in which different groups of teachers in different schools have set about tackling the problems created and seizing the opportunities offered by the introduction of mixed-ability classes. The first aim of this new edition, therefore, is to modify what was said before in the light of this new experience and to incorporate some of it into a more up-to-date discussion of the practicalities of mixed-ability teaching.

The second aim is to provide a more substantial theoretical underpinning to this discussion. All purposive activities must have a theoretical basis. It is not possible either logically of practically to undertake any purposive action without some understanding of the reasons why one is doing it and the principles and intentions that lie behind it. It is this simple truth about purposive action that makes that rejection of theory that some few teachers still boast of a complete and literal nonsense. In the present context there are two interrelated aspects to this need for a theoretical basis. In the first place, teachers need an understanding of the fundamental principles of mixed-ability teaching if their practice is to be successful. Such an understanding is essential if they are to know *how* to set about educating their pupils in mixed-ability classes. However, secondly, they need also and for the same reason to understand *why* they are being asked to educate their pupils in such classes. As we have just said, successful teaching also requires the kind of commitment to the system in which one is working that can only come from a conviction that it is a right system.

The first edition of this book attempted, albeit briefly, the former; it did not essay the latter, since at that time it was felt that most of those teachers who were involved in mixed-ability teaching were convinced of its value and needed guidance only on its principles and practicalities. It was also felt that too theoretical an approach would be likely to deter teachers from further interest. Now, however, perhaps because this kind of dramatic change in their working conditions has encouraged in many teachers a more thoughtful approach to their work, it seems that teachers do have more respect for what educational theory can offer them. Also, in the light of the reaction against mixed-ability grouping described above, it is clear that the question of its pros and cons must be faced.

It is the second aim of this new edition, therefore, which is reflected in the change of title. For it is now the intention to offer some thoughts on the question of *why* we should group pupils by criteria other than their measured intellectual ability as well as *how* we should set about educating them once so grouped. This question I have tried to face both explicitly in a new opening chapter that attempts to offer a rationale for mixed-ability grouping and implicitly by trying to discuss the practicalities of mixed-ability teaching throughout with this rationale in mind. In a sense the whole book has been restructured in an attempt to develop this kind of rationale while at the same time adhering to the original intention of assisting teachers with the practical day-to-day business of teaching pupils in such classes.

It is my hope that in doing so I have succeeded in offering not only a rationale for this particular aspect of educational innovation but also a basis of professional confidence on which teachers will feel able to continue to make modifications in both the matter and the manner of their educational provision. For without this kind of continuous change, based on the confident professional judgement of its practitioners, education will become a static enterprise and, as with all other human activities, once that stage has been reached, all life will have gone.

A. V. Kelly
London, January 1978

CHAPTER 1

A RATIONALE FOR MIXED-ABILITY GROUPING

The system of grouping pupils by both age and ability so that they could 'stream' through the school in the channels appropriate to them was introduced into the new primary and secondary schools in the United Kingdom in the 1930s and was still accepted as of self-evident validity, at least in the secondary schools, by all but a few intrepid innovators until well into the 1960s. Indeed a survey of the attitudes of primary school teachers undertaken in the late 1950s revealed that of the 172 primary teachers questioned on the most appropriate method of grouping pupils on entry to primary schools every one felt it necessary to do this on the basis of measured attainment and ability (Daniels 1961a) and, while it could be argued that the approach of that study invited such a response, it is also true to say that this reflected the practice of the time.

However, there were schools at that time which had already abandoned streaming. A few of these were secondary modern schools but most were primary schools. This is clear from the Ministry of Education report on primary education published in 1959, which, as Brian Jackson reminds us (1964), described streaming as 'peculiar to our own day' and suggested that it was not impossible to teach classes of mixed ability. Its words are worth quoting not least because of the time at which they were written.

> One of the most remarkable developments in teachers' skill over the last decade or so has been that of educating in one class children of very different abilities. They do this by arranging the environment in the classroom and school so that the children learn a great deal for themselves, either individually or in small groups. The teacher comes to know when to teach groups and when to teach the class or an individual, and in time he knows at what pace and in what ways the different groups of children learn best. To acquire this is no mean achievement, though to many who have an understanding of children and a fertile inventiveness it appears to come easily (p. 70).

Perhaps this report stimulated the development, for, at the beginning of the 1970s, it was difficult to find a primary school that grouped its pupils into classes on the basis of ability — at least in any overt manner — and a survey of 1000 comprehensive secondary schools undertaken by the National Foundation for Educational Research in the early 1970s indicated that 50% of those schools had unstreamed first years, 37% unstreamed second years and 24% unstreamed classes in year three — figures which probably represent the growth of this movement or differing views about the age-ranges for which it is appropriate rather than a decline of enthusiasm for it.

Several points can and should be made about this development and particularly about its rapidity. In the first place, it must be acknowledged that administrative decisions have again played their part in bringing this about. The abolition of the 11+ and of selective secondary education has certainly relieved some of the pressures on junior schools to conform to previous patterns and has made change possible, although not, of course, in itself requiring the abolition of streaming. The comprehensivization of two-form or three-form entry grammar schools, however, has resulted in the emergence of nonselective secondary schools which cannot avoid organizing themselves on some kind of mixed-ability pattern. To admit to a school 60-90 pupils on a nonselective basis, covering the full spread of intellectual attainment, means on present teacher-pupil ratios that one has two or three classes of a very wide spread of ability, no matter what basis is chosen for the grouping — an experience, incidentally, not unfamiliar to those people who have taught in secondary modern schools of similar size in areas which offered very few grammar school places.

In spite of the fact, however, that administrative decisions have forced many schools into this position, there has been very little evidence of any lead being given to schools by those responsible for these decisions or those responsible for advising them on their development. Little help has been forthcoming either on the question of the desirability or undesirability of certain forms of grouping or on that of the ways in which the problems created by these changes might be tackled or the opportunities used. A pamphlet produced by the Ministry of Education as early as 1945, *The Nation's Schools*, did question the desirability of streaming and, as we have seen, the report on primary education of 1959 did draw attention to some of the advantages of a mixed-ability form of grouping in primary schools, but beyond that there has been no kind of official lead offered at any level. The recent survey of mixed-ability teaching at secondary level within the Inner London Education Authority, for example, undertaken by its inspectorate, has resulted in the production of a report that is remarkable for its neutrality and lack of any kind of constructive advice for the teachers who find themselves actually involved in the practicalities of mixed-ability classes.

In spite of this absence of guidance, however, many schools have gone ahead with plans for unstreaming and the introduction of some kind of mixed-ability organization, even when this has not been forced upon them by administrative change. In some cases, one must admit that this appears to have been prompted either by a desire to leap onto any bandwagon or by an attempt to clutch at any straw that seemed to offer the hope of providing a panacea for the obvious ills that have beset many secondary schools in recent times — truancy and other behaviour problems in particular. It is also clear, however, that this movement has been prompted equally often, perhaps more often, by an awareness on the part of many teachers and headteachers of some of the inadequacies of streaming, a recognition that some change in education is necessary if the practice of schools is to keep pace with the movement of the times and a zeal for a system that seemed to be based on a sounder interpretation of the egalitarian philosophy of the 1944 Education Act.

More recently, however, one has detected something of a reaction to this movement. Certain events that have caught the public eye, such as the affair of the William Tyndale school, have increased the suspicion of anything that might be described as 'progressive' in education, although why the term 'progressive' should be given this pejorative meaning is difficult to explain. These events have led to a national public debate on education, to the establishment by the Department of Education and Science of an Assessment of Performance Unit, to demands for a reduction of teacher autonomy through a monitoring of standards and perhaps the introduction of a common core curriculum, to an attempt to restore some kind of authority to the inspectorate at both local and central government levels and, as the sad but inevitable result of all of this, to a loss of confidence on the part of many teachers in their ability to promote the right kinds of educational innovation.

As far as mixed-ability teaching is concerned, this general trend has been reflected in two ways. Firstly, as was suggested earlier, it has led to a slowing down of what was at one stage an indecently and perhaps unwisely rapid move in this direction, which certainly resulted in the introduction of some ill-considered schemes which did the notion of mixed-ability teaching more harm than good. Secondly, it has meant that there is now a quite proper demand for a more carefully thought out and positive justification of this form of grouping. A rationale is needed because teachers need both to believe in a system and to understand it if they are to make it work and such belief and understanding now need to be established or restored.

There is ample evidence that any system of teaching will work only to the extent that the teachers responsible for implementing it accept it and are committed to its value and, indeed, its values. It is equally clear that no such commitment is possible in any acceptably professional sense unless it derives from a full understanding of the reasons, the justification and the implications of any scheme of this kind. Nor is it only commitment that requires this kind of understanding; a proper practice will require it too, since one cannot do any job that requires the exercise of judgement properly without the theoretical understanding that makes the exercise of that judgement possible. Thus any attempt to assist teachers to improve their practice must proceed by endeavouring to help them to develop their thinking and their understanding as well as offering them advice of a more practical kind. It is in this direction that a proper relationship between the theory and the practice of education is to be found.

For both of these reasons, then, a rationale for mixed-ability teaching is now needed and it is the aim of this new opening chapter to provide the beginning of such a rationale.

A preliminary point needs to be made, however, before we embark on this, and that is that there are enormous difficulties in achieving any kind of adequate or precise definition of mixed-ability teaching (Reid

1977). There is no one thing that we can point to as offering us the paradigm here. Indeed, I hope to argue eventually that this is the exact source of the strength of this system, that it offers the kind of flexibility that allows for many of the individual adjustments, modifications, and interpretations that anything that is to be called education in the full sense must evince. However, that very flexibility does present us with problems of definition.

For example, we can gain no agreement on the meaning of 'ability' whether we want to use that notion to group pupils by ability or to mix the abilities. Nor is it possible to find agreement on the relationship of ability to attainment, a concept that includes such additional factors as motivation, attitude, specific skills and so on. Thus mixed ability is often confused with mixed skills, mixed attitude, mixed motivation and mixed attainment generally. This is reflected both in the different interpretations of schools as to what mixed-ability grouping is and in the many different methods they adopt when setting about it. Thus some schools will describe a system of 'broad banding' as mixed-ability grouping; some will have mixed-ability groupings only for certain subjects; some schools, including many primary schools, will allow mixed-ability classes to be taught by grouping pupils by ability or streaming them within the class; while others will regard any or all of these practices as not qualifying for the description of mixed-ability teaching at all.

The methods by which schools create their mixed-ability classes will also vary enormously — a deliberate mixing of abilities based on IQ scores, attainment tests or primary teachers' assessments (i.e., what those teachers think is the level of motivation, attitudes to work and so on of their pupils); friendship or neighbourhood groupings or groupings based on primary schools of origin; or totally random, perhaps alphabetical, grouping. There are as many different ways of creating a mixed-ability class as there are definitions of what is meant by mixed ability. It will be clear too that each of these methods and each of these interpretations will produce different kinds of 'mixed-ability' class. All of this has become apparent quite early in the experience of the current NFER Mixed Ability Teaching Project (Reid 1977).

These problems of definition, then, bedevil us from the outset, but I hope to show ultimately that the only viable definition of mixed-ability grouping is one that allows for flexibility and adaptation to meet precisely these different local conditions, different views taken by different schools and groups of teachers and, above all, the many different circumstances and contexts of the proper education of each individual pupil. I hope also to show not only that mixed-ability classes should be created in the light of individual school situations but also that the mixed-ability class should not be seen as a fixed and permanent grouping for all the many purposes a school will have but should be regarded merely as providing a base upon which many other kinds of grouping can be built or from which they can be developed.

It is against this rather confused backcloth, therefore, that we must endeavour to produce a rationale for mixed-ability grouping. To do so there are two kinds of argument to be produced. In the first place, there is what one might call the negative argument, that which is based on the disadvantages that are seen in the system of streaming, arguments which suggest that streaming has not proved to be a successful system and that something has to be found to replace it. These have tended to be the dominant arguments in most recent discussions and they have often led to something called 'unstreaming' rather than mixed-ability grouping. Secondly, there are what one might call the positive arguments for a mixing of abilities of some kind, the kinds of argument that claim that there are good reasons for saying that what should replace streaming is mixed-ability classes or perhaps the grouping of pupils without reference to measured ability. These are the arguments that must be developed if a real rationale for mixed-ability grouping is to be established.

I have tried to show so far, then, that a rationale for mixed-ability grouping is now needed, that the production of such a rationale is be-devilled by problems of definition and variety of practice and that if it is to be achieved it must be approached from two directions, from a consideration of the criticisms that have been levelled at the system of streaming it is designed to replace and from more positive arguments for replacing that system with mixed-ability classes. Each of these approaches we must now consider in turn.

A negative rationale — the case against streaming

Any evaluation of streaming must begin with a consideration of why it was introduced in the first instance, since one must look very closely at the assumptions upon which it was, and in some cases still is, based. There were really two broad reasons for the introduction of streaming. The first of these was again an administrative reason and we must never lose sight of the extent to which administrative decisions govern educa-tional procedures. It arose from the idea, first made public in the Hadow Report of 1926, that everyone should have access to secondary education.

The system of grouping favoured in this country in the early part of this century and, indeed, from the introduction of state education on any real scale was that of standards, a system of grouping by ability but not by age. Under this system, the origin of which is closely associated with the method of 'payment by results', children moved from one standard to the next when they were deemed to have reached a certain level of attainment, those who did not reach the necessary level remaining where they were until they either improved sufficiently or reached an age at which they could leave school. When places in grammar schools were made available to bright pupils, the effect on this system, leaving aside the increase in the attention given by teachers to those pupils who might be able with coaching to secure such places, was no more than to remove the brighter pupils from the higher standards. While all other pupils remained in an all-age school, the system of standards sufficed.

Streaming as such arrived, as we have just suggested, with the attempt to establish secondary education for all which began with the Hadow Report on the Education of the Adolescent in 1926. The 'decapitation' of the all-age elementary school which the establishment of secondary education for all entailed resulted in the disruption of the standards system and the emergence of the dull and backward in the new primary and secondary schools. It also placed an increased emphasis on the selective procedures operated at 11+, since these now were to be used for the allocation of all pupils to the appropriate form of secondary education, not merely the selection of a few very able pupils for grammar school places, and thus rendered an emphasis on intellectual development in the teaching and organization of the new primary schools inevitable.

There was a second factor which contributed to the disappearance of standards and the emergence of streaming. Psychologists had for some time been stressing the need to take account of age differences in grouping children for teaching purposes and the undesirability of having too great a mixture of ages in one class, claiming that the grouping of children according to chronological age produced fewer learning problems. At the same time the work of psychologists such as Binet on intelligence and Burt on backwardness encouraged the view that ability could be measured and used as a basis for grouping and that less able pupils in particular needed to be given special treatment, a view that has continued to form the main plank in the case for streaming for many people since.

These factors together suggested that in the new schools standards should be replaced by streams and in fact a 'triple track system of organisation' of pupils into A, B, and C classes was explicitly advocated by the Hadow Report on the Primary School in 1931. In 1937 this official view was reiterated in the Board of Education's Handbook of Suggestions for Teachers and the same system was suggested as appropriate for the multi-lateral school in the Spens Report on Secondary Education in 1938. The ideology behind this was of course, the ideology of the day which led to the organization of secondary education itself as well as the grouping of pupils in classes along selective lines.

Many of the assumptions upon which this system is based have subsequently come to be questioned. In the first place, this whole episode in the history of the English education system must cause us to question the validity of any educational decision that is made largely on the basis of the evidence currently being profferred by exponents of psychology or, indeed, of any other single discipline. Education is not applied psychology nor applied sociology nor the application of any other such area of study; it is a far more complex undertaking and involves many different and sometimes conflicting considerations. We should always, therefore, look very closely at administrative decisions that appear to be based on the recommendations of one group of academics, not least because we know that academic fashions change or rather that all bodies of 'knowledge', especially in the highly problematic area of the human sciences, are subject to constant modification, so that all administrative decisions made on this kind of basis will later need to be unscrambled if they are not to perpetuate what will by then have become outmoded theories.

This point of view is certainly borne out by the case before us here. For the theories upon which the psychologists of the time based their recommendations no longer enjoy general acceptance today and this is where the case against streaming begins. Many people no longer accept the rather simplistic notion of intelligence which was clearly fundamental to the introduction of streaming. There is now no general acceptance of the view that intelligence is a kind of faculty nor that it is one whose efficiency is largely determined by heredity. Such a 'psychometric' view has been largely replaced in the eyes of many theorists by a developmental view which sees intelligence more as cognitive growth or intellectual development and as a result subject to constant modification by environmental factors or at least as being as open to the influences of the environment as to the determination of birth. No matter how strong the hereditary factors are, therefore, we can never know what they are, so that we cannot put limits on the scope of environmental influence on the intellectual development of any pupil.

This in turn has led to a questioning of the validity of intelligence testing and especially of its predictive validity. The architects of streaming believed that what a properly constructed IQ test would tell us was not just the level of intelligent performance being revealed by the child at the time of testing but also his intellectual capacity and potential. While this may have some validity in the case of the pupil who scores highly on such a test, there are no grounds for assuming that those who do not score highly could never do so, given the right kind of intellectual stimulus. It was precisely this consideration that led Lord Boyle, as the current Minister of State for Education, to preface the Newsom Report in 1963 with the words 'The essential point is that all children should have an equal opportunity of acquiring intelligence.' It is the possibility of such acquisition that offers scope to the teacher. Streaming was based on a denial of that possibility and the assumptions behind this have been roundly questioned.

A further aspect of this is the concept of 'general ability' with which this theory of intelligence was also associated. For it was assumed that what was revealed by an IQ test would have significance for all kinds of intellectual activity — an assumption that every teacher must have had frequent cause to question — and that any grouping of pupils on this kind of evidence would give suitably homogeneous groupings for all the purposes of the school. But, whatever view we take of intelligence, most of us nowadays would appreciate that it manifests itself differently in different spheres and the notion of something called 'general ability', which is fundamental to streaming, is highly questionable in the light of current psychological thinking.

Lastly, even to attempt to produce classes that are homogeneous in terms of measured intellectual ability is to make certain assumptions about the teaching methods to be adopted. It assumes a largely didactic, class-teaching approach and thus adopts a rather unsophisticated model of teaching, the model of a teacher teaching a class the same material in the same way at the same pace and at the same time. But we are now beginning to realize how unsophisticated and inadequate this model is and that it is not the case that the only or even the best way to learn is by being taught. Most of our primary schools realized this some time ago; many secondary schools are coming to appreciate it; and even some institutions of further and higher education are beginning to recognize the advantages of other approaches. Our understanding of teaching and learning, therefore, has progressed considerably in the intervening years and we are now aware that, although this approach has its merits — even with mixed-ability classes — it also has its limitations and that the most effective methods of teaching are those that recognize the need not so much to teach as to facilitate learning and thus acknowledge the advantages that 'less formal', 'heuristic', 'individualized' methods can offer. If we are to adopt such methods, there is less need to attempt the impossible task of achieving intellectual homogeneity in our teaching groups. Again, since this assumption has been questioned, one must reconsider the system of grouping that was based on it.

Thus the case against streaming begins from a questioning of the main assumptions behind it simply because our thinking about education has progressed over the forty or fifty years since streaming was introduced. In particular, we have come to question its assumptions about the nature of intelligence, the validity of intelligence testing, and the most efficient ways of setting about the task of teaching children.

The second major feature of this negative case for mixed-ability teaching is based on the evidence of the effects of streaming that has mounted as a number of different studies have looked at various aspects of our educational system and, more recently, at streaming itself. There are two parts to this which it might be helpful to consider separately. First of all, there is the evidence of the extent to which streaming can be seen as a contributory factor to that disturbing feature of the educational system of England and Wales, the wastage of talent that we first became aware of in the 1950s. And, secondly, there is the evidence of the effects of streaming on the progress of individuals which has emerged from those studies that have been undertaken of streaming itself. The national investment aspect of the education system and the provision of adequate and suitable educational opportunities for every pupil are equally important goals, as the Crowther Report reminded us, and we must consider the apparent effects of streaming in the light of both of these considerations.

An alarming wastage of talent from our schools was revealed by research that preceded the publication of the report on Early Leaving in 1954 and the Crowther Report itself in 1959 and this evidence was supported by other studies such as that of J. W. B. Douglas (1964). It became abundantly clear that many pupils whose ability was demonstrably high were leaving school with little to show for the time they had spent there in terms of qualifications that would enhance the contributions they could make to society or their own prospects of living a full and satisfying life. In particular, of course, these were pupils of 'working-class' origins and the production of this evidence sparked off a whole series of studies designed to establish the causes of this wastage.

Inevitably the first point of attack for these studies was the homes of these pupils and the search began for factors there that would explain what was happening and provide some clues as to how it might be corrected. Many such factors were, of course, revealed, but what also emerged very quickly was that the schools themselves were not only failing to counteract these factors but were actually doing a good deal to aggravate them as well as making their own contributions to promoting this wastage.

Recently, the curriculum itself has come in for a lot of close scrutiny in this respect; we have begun to think that perhaps the content of what we are offering and the way in which we are offering it may be factors to be considered. But the first area of exploration was the selective procedures that were in use for allocating pupils to different kinds of school and subsequently for allocating them to classes within the schools. The processes of selection generally came under attack and this has led to the abolition of the use of selective procedures in the assigning of pupils to secondary schools as well as to that abandonment of streaming in many schools that we referred to earlier. The parallels between the comprehensivization of secondary education and the introduction of mixed-ability classes are very close since fundamentally both cases are based on the evidence of the effects of selection and on a growing conviction that it is impossible to devise procedures that will enable us to select with any acceptable degree of accuracy.

For it has become quite clear that wherever selective procedures are used they result in many mistakes being made. In part this is due to that lack of predictive validity for the tests used that we noted earlier. To measure an individual's attainment at the age of seven or eleven or sixteen or eighteen is in most cases to obtain only the flimsiest evidence for predicting his future academic achievements.

Secondly, however, it is also due to the unavoidable fact that many irrelevant factors enter into our decisions as to what stream a pupil should go into or what kind of secondary education he should have. Thus it became apparent that entry to a grammar school or to the 'A' stream depended as much on such things as the area of the country a child happened to live in, the month of the year in which he was born, the childhood illnesses that kept him away from the infant school and many other factors as irrelevant to educational decisions as those (Jackson 1964).

It also became apparent that once placed in a 'B' or 'C' or 'D' stream or in a secondary modern school one's self-image would take a dive and one would begin to see oneself as intellectually inferior. It appeared too that the gaps in attainment between the streams became progressively wider (Douglas 1964). There was also some evidence to suggest that it was the 'bright' pupil in the 'B' stream whose performance deteriorated most in these circumstances and this fitted exactly with the evidence of wastage that the Early Leaving and Crowther Reports had produced.

One of the explanations of this phenomenon that began to emerge was that of teacher expectation. If we put a label such as 'A', 'B', 'C', 'D', grammar, technical, or modern on children, teachers will inevitably treat them accordingly, developing those curricula and levels of work that they deem to be appropriate to them, never expecting more from any of them and, therefore, never getting anything more (Pidgeon 1970; Downey 1977).

Furthermore, the effects of this were clearly aggravated by the fact that in most schools teachers were 'streamed' as well as the pupils. 'A'-stream pupils, therefore, usually had the added advantage of working with the best, 'A-stream' teachers, while those pupils placed in the other streams had the corresponding disadvantage of working with the less inspired and inspiring members of staff — a practice that had the same implications for the morale of teachers as we are claiming streaming has for that of pupils. This practice is well illustrated by Arthur Young's comments on the system that was in use before the introduction of mixed-ability grouping at Northcliffe Community High School (1975, pp. 31-32).

> There were other aspects of 'ABCD' which led to inevitable difficulties. The matching disciplinary patterns were also self-fulfilling prophecies. 'A' to 'D' in ability synchronised almost exactly with 'A' to 'D' in behaviour. Like most Headmasters, I also found it necessary to match staff to these requirements. The most imaginative and sensitive teachers for the 'A's', almost anyone would be able to manage the 'B's', the toughest and most insensitive for the 'C's' and thank God for volunteers — very often indeed the most dedicated ones — for the 'D's'.

This resulted therefore, in a low level of transfer both between schools and between streams. Errors of allocation were always accepted as un- avoidable both in selection at 11+ and in the allocation of pupils to streams, but it was always assumed that these would later be put right. However, the long-stop device of the 13+ never worked — Jackson (1964) revealed that the national rate of transfer to grammar schools from others at age thirteen was 1% — and the number of pupils who were reassigned to other classes after being incorrectly streamed was always very low (Daniels 1961a; Douglas 1964; Barker-Lunn 1970). Vernon (1960) suggested that about one-third of primary school children should change from the stream they were first placed in and about 10% would need to change each year; and the teachers involved in Daniels' study (1961a) belived that a high rate of transfer between streams would be necessary to correct errors in initial selection — about 90% of them expected to have to transfer between 10% and 30%, with a heavy concentration around 20%. But Daniels found that in the schools he studied the figure was nearer 5% or 6%. Douglas too, looking at his large sample of children in their primary schools, said, 'It was rare for children to change stream. Over the whole three-year period the annual rate of transfer was 2.3% and approximately the same number moved up as down. On this showing the system of streaming by ability is more rigid than is generally realised' (p. 113).

Thus it became abundantly clear that, once streamed, pupils were on a predetermined course with little likelihood of being deflected from it. They were victims of an 'inbuilt finality of judgment, so hard to over- master' (Jackson 1964, p. 124), and, as a result, streaming appeared as a major factor in hindering the achievement of equality of educational opportunity.

Selection procedures, therefore, do not and cannot work either efficiently or fairly. For this reason they are now no longer used in the allocation pupils to secondary schools in most areas of the country and it is one of the strongest arguments against streaming within schools that they cannot be used there without resulting in similar inaccuracies and injustices.

This, then, was the source of the initial impetus away from streaming. Inevitably it led to a number of studies into the effects of streaming itself and the relative merits of streaming and unstreaming, and it is to the findings of these direct studies of streaming that we must turn for the last element in our case against streaming.

The first thing one has to say about this evidence is that it is in-conclusive with regard to academic or intellectual attainment. We have already referred to the evidence that the effect of streaming is to widen the gaps between the streams and the spread of ability generally (Douglas 1964; Jackson 1964). There is some evidence that mixed-ability grouping has the effect of reducing that spread, especially by raising the level of attainment of those who would have been placed in the lower streams (Daniels 1961b). This latter evidence, however, is based only on research into junior schools and there are too many variables, preeminent among which is the quality of the teaching, for us to regard it as in any way conclusive.

But this inconclusiveness is in itself interesting enough. For if there is any conclusion to be drawn from it, it is that streaming does not bring the success that it was claimed it would bring even in the purely academic or cognitive sphere; there is no evidence that able pupils do in fact achieve more highly in streamed classes or less highly in mixed-ability classes nor does it seem that the less able do better when given special attention in streamed classes, since, in fact, they seem to do rather better when working side by side with pupils of all abilities. Slightly more conclusive support for each of these claims has emerged from a more recent study of a large comprehensive school at Banbury. 'There was little direct evidence to suggest that high ability pupils were achieving markedly differently in the two systems. For low-ability children there was a significant gain in the mixed-ability system compared with the streamed' (Newbold 1977).

Thus it has been claimed that the decision whether to stream or not must be taken on other grounds (Barker-Lunn 1970). Preeminent among these other grounds are considerations of the impact of streaming on the social development of children. Here the evidence is more clearcut both in terms of the negative effects of streaming and the positive effects of unstreaming, since every study undertaken has revealed the same thing. Elsa Ferri, for example, examining for the NFER the long-term effects of streaming and nonstreaming in the junior school on the subsequent personality, social and intellectual development of pupils in the secondary school, said, 'Comparisons between the streamed and non-streamed approaches showed that while pupils' academic progress was largely unaffected by the type of school which they attended, their emotional and social development was influenced, both by the school's organisational policy and by the approach of their teaching' (Ferri 1972, p. 71).

Teachers have long been aware of the behavioural problems they were presented with by 'C' and 'D' streams, particularly in the upper reaches of the secondary school. Recent work has now confirmed that a very different kind of social development goes on in these classes from that which most teachers would want to promote. David Hargreaves (1967), for example, found that 'A'-stream boys were keen on school activities, treated the staff with respect and were content to meet the demands made of them by the school. 'D'-stream boys, on the other hand, were uninterested and often completely hostile to the school and their teachers, the result of a negative self-image often reinforced by the attitudes of the teachers towards them. Thus there grew up an antagonism between the two groups of pupils and there developed academic and delinquescent subcultures at loggerheads with each other. Colin Lacey's study (1970) revealed much the same phenomena in a selective boys school — differences in attitude towards the school and its values, different levels of participation in extracurricular activities and a polarization of the different groups of pupils.

Pupils allocated or relegated to lower streams, therefore, appear to lose all sense of identification with the aims and purposes of the school; they see themselves as not valued by the school, not wanted by the school, so that they develop their own culture which is based on opposition to school rather than any collaboration with it; they come to place their faith in and take their strength from the peer group, so that a delinquescent subculture emerges, the prime goal of which is to reject and even to subvert the values of the school rather than to embrace them. Thus there results a very strange kind of social education — if one can even call it that — for pupils placed in these lower streams.

This is reflected especially in the response of pupils to the extracurricular offerings the school makes. Again Arthur Young's comments illustrate this well (1975, p. 32).

> We also had the evidence of the Duke of Edinburgh Awards which puzzled us greatly, for here was a scheme organised in such a way that it should appeal to young people of all abilities — yet it was not found to work in this way for us. We were finding that although we started with groups from all streams, most of the 'C' and 'D' streamers were dropping out after only a few weeks. Another case of the disadvantaged being further disadvantaged. Even an analysis of our school holiday journeys showed that most of the candidates came from the 'A' and 'B' streams. It could not be a social class correlation in a school like Northcliffe, because nearly all of them came from the same socio-economic group.

Conversely, few of those teachers and headteachers who have abandoned streaming would wish to deny that in social and behavioural terms there are immediate and lasting gains. This is confirmed by the NFER study of streaming in the primary school (Barker-Lunn 1970), which unearthed clear evidence of improved pupil-pupil and teacher-pupil relationships, greater involvement by all pupils in the life of the school, as measured by such things as the level of participation in extracurricular activities, and in general a healthier social climate in unstreamed primary schools.

Moral and social education are as much a function of the way in which we organize our schools as of any attempts we make at direct teaching in these areas. The values that are implicit in the way in which we set about our business as teachers are much more likely to be 'caught' by pupils than are those values we hope will be taught by more formal methods. And so we must accept that this evidence of the relative effects of streaming and unstreaming in this area is of great significance. Thus again a positive argument for mixed-ability grouping emerges from an awareness of some of the effects of streaming.

Lastly, we must note the implications of streaming and nonstreaming for the development of children's ability to think divergently. It is clearly important that in a rapidly changing society we should endeavour to promote this kind of ability in our pupils. It is clear too that this is another way in which our view of education has changed since streaming was first introduced, for then we were unaware — perhaps understandably — of the need for this. There is now a certain amount of evidence to suggest that the development of divergent thinking is more effectively encouraged in the less formal atmosphere of nonstreamed schools than in the more formal atmosphere of those that stream (Haddon and Lytton 1968; Ferri 1972). Nor is it to be wondered at if pupils find it easier to 'chance their arm', in the way that they must if they are to be encouraged to offer their own solutions to problems, in a situation which is noncompetitive and where failure will not cause one to slip down any competitive educational ladder.

The negative case for mixed-ability teaching, then, is based on three kinds of consideration — firstly, a questioning of the assumptions on which streaming is founded, secondly, the growing evidence of the contribution of streaming to many of the ills of education that have been very clearly delineated over the last few decades, and, thirdly, the more recent evidence from studies of streaming itself of the effects it has on certain aspects of the upbringing of children and particularly on their social development.

All of this adds up to a case for some kind of change. Streaming has been shown to be an inefficient form of organizing teaching and we have to change to something else if we want our schools to be efficient. Indeed, some kind of change would seem to be essential in a society that is itself undergoing rapid change on all fronts. Whether that change should be to a mixed-ability form of teaching we must now consider as we attempt to establish a rationale for mixed-ability teaching by looking at the kinds of positive argument that can be produced to support it.

A positive rationale — the case for mixed-ability grouping

If we accept that schools should change to meet the changes in society and that some new system is needed to enable schools to do so, one kind of positive justification for mixed-ability grouping must be sought in a consideration of how far it enables schools to meet the new demands being made of them by changes in the social setting in which they operate.

It is clear to begin with that the technological developments of the last half century make it vital to educate everyone to the maximum of his or her potential. A highly complex technological society is voracious of educated people at all levels. When streaming was introduced we did not need to educate everyone in this way; we needed only to tap the obvious talent that we had. Thus, as we have already indicated, streaming is very wasteful in this way. A positive rationale for mixed-ability teaching will, therefore, need to be based in part on evidence that it is a more efficient, if not the most efficient, system when judged from this point of view. A lot of people would want to argue that educating all pupils together is likely to lead us closer to this goal. Certainly, separating pupils out has singularly failed to do so.

Secondly, it would be argued that we need to educate all to mix with all. We need to produce people who can work with one another. In other words we do not want to produce a society in which there is a very clear segregation of classes or groups of people, whatever that segregation is based on. Again, it would seem probable that educating everyone together is more likely to bring about such a state of affairs than dividing them up.

Thirdly, technological change also requires that people be adaptable and able to adjust to the continuous change, both technological and social, that they will constantly meet. They will need what the Crowther Report described as a general mechanical ability rather than a more specific kind of training; they will need to be flexible and adaptable and able to think for themselves; and again many would want to argue that being taught in mixed-ability classes will contribute to this.

However, a much stronger case for mixed-ability grouping is to be found in an examination of the extent to which it reflects wider changes in the ideology of society. For technological change has led to social change, to changes in the value system or systems of society. Views on most aspects of social living have changed over the years since streaming was introduced and education is no exception to this. To recognize the errors and inadequacies of streaming is to acknowledge the weight of these changes. It is to be committed so something more, to a different set of expectations, a different system of values, a different ideology from that which prevailed when streaming was first introduced. The case for mixed-ability grouping must, therefore, be based on an assessment of its ability to reflect and respond to these changed social conditions. What are the chief ingredients of this new ideology? There are several tentative answers that I want to suggest to that question, several features of this new system of values that I want to try to tease out.

Preeminent among these features would seem to be an acceptance of a social and educational egalitarianism of a 'democratic' kind. We have left behind the meritocratic social philosophy upon which the introduction of streaming was based. The 1944 Education Act obliged us to educate every child according to his age, aptitude and ability and, although that directive is capable of a wide variety of interpretations, as will be clear from the most cursory survey of the different attempts that have been made to implement it, the main thrust of the act was towards offering something of educational value to all pupils and not merely giving each the opportunity to achieve on merit an education defined in largely academic terms. What this entails was better expressed by the Crowther Report which spoke of education as the right of every child regardless of any return that society as a whole may or may not receive from the money it has invested in its education system. It is only recently that we have come to accept that this ideal is not attainable by devising two or three different kinds of education and hoping that there will be parity between them. It can only be attained by abandoning any hierarchical view we once held of the relative status of different kinds of knowledge or activity and adopting an approach to education which recognizes that it should be designed to offer everyone something of value defined in terms of his own individual requirements. In other words, we have a new view of education not as a privilege, not as something we offer to those who can give us some return for it, but as something that is the right of everybody and which must be provided for everybody in the fullest possible sense — education for the individual and not only for society. Educational provision, then, is to be justified because it offers the individual something worth having and not merely because society can get some return from it — a point that we would do well to keep in mind as we evaluate the current pronouncements of politicians and others about education. Such an ideal can only be achieved by a nonselective form of organization and an acceptance of this kind of view is one feature of the changed thinking that has led to the advent of mixed-ability groupings.

The same considerations lead us also to recognize that we can no longer regard education as a kind of competitive game, a race with a limited number of prizes, since to do so will not be consistent with the idea of treating every child as of equal value and every child's achievement as equally acceptable. To say this, of course, is emphatically not to say that the intention should be to conceal the differences, even inequalities, that can be discerned between people. It is to say, however, that the differences and inequalities that exist between people's ability to read, to understand mathematics, to write French proses and so on should not be regarded as though they were indications of fundamental differences between them as human beings, as though a lack of ability in any or all of these areas was evidence of membership of a qualitatively inferior race. We should by now have left behind the Platonic classification of men into groups of gold, silver, and bronze defined according to their intellectual capabilities.

This also implies that we can no longer pretend that any educational goal is achievable by the promotion of competition between pupils. Only if we define education in terms of cognitive achievements of the most unsophisticated kind can such a view be tenable. For if we accept the definitions of those who, for example, suggest that to be educated is to have come to value certain kinds of pursuit for their own sake, then pursuing these things for the purely instrumental reason of getting high marks or, worse, of getting better marks than other children in the class or if getting into the 'A' stream rather than the 'D' stream would seem not only unlikely to promote such an attitude but even likely positively to discourage it. It has never failed to puzzle me throughout my years in the teaching profession to find so many teachers whose main reason for teaching is their love of their subject but who attempt to get their pupils to share that love by whipping them into a frenzy of competition with each other. Nor is it any kind of argument to say that men are naturally competitive and that society as a result is a hive of competition. This could never constitute an acceptable argument for promoting anything in education. If it could, then a similar argument from those who believe that man's behaviour is motivated primarily by sex and aggression would have some very interesting implications for curriculum planning. We must accept, therefore, that mixed-ability grouping is prompted in part by, and held by many, to reflect a view of man as a social animal rather than as a competitive predator, or at least the view that education should not be designed to promote his competitive instincts, and the allied conviction that education in the true sense cannot be forwarded by such methods.

This brings us naturally to the third point I want to make about what mixed-ability grouping means in terms of our view of man, society and education. For it is certainly also a direct result of the fact that the view we now have of education itself is very different from that which prevailed when streaming was first spawned. Then education was seen as almost entirely a matter of cognitive development, sometimes, as we have just suggested, of a very limited kind. If it is only cognitive development, intellectual attainment, that we are trying to foster, then there may be some point in grouping children according to what appears to be their capacity for assimilating this kind of offering. This is especially true, as we saw earlier, if one also believes that this kind of goal is best attained by teaching of a largely didactic kind, since again it may help to have a group of children who are likely to digest what is fed to them at roughly the same rate.

However, a glance at the curriculum of any modern school and a comparison of that curriculum with those in vogue forty or fifty years ago will quickly reveal that many changes have taken place as a result of a rethinking of what education is all about. The briefest observation of the practice of most schools will also demonstrate the truth of what I am claiming. For we now take a much wider view of education and our aims and practices reflect this. We now accept far more readily the importance of affective goals; we have taken seriously the claims of those who have advised us to endeavour to promote creativity; we believe that the development of children's feelings is our concern; we regard their emotional development as of some significance; and we have come to recognize that social and moral education are equally important parts of the school's role.

Such sweeping changes in our educational ideology must imply changes in the organizational structures we design for the practice of education and the move to mixed-ability groupings also reflects this shift in our view of what education is fundamentally about.

Furthermore, we have realized that many of these dimensions of education are actually affected as much by the ways in which we organize our schools as by the overt provision that we make in our time-tables, syllabuses and lessons; in short we have come to recognize the significance of what some have called the 'hidden' curriculum. For certainly social and moral learning and emotional development are as much a function of the organization of the school as of any positive attempts we make to promote them and we have already referred to the particular kinds of social learning and emotional development that are forwarded by streaming. Nor, as we have seen already, can we expect much creativity to be shown by pupils in a competitive and hierarchically ordered setting.

The introduction of mixed-ability groupings, therefore, also reflects this wider conception of what education is. Certainly, the kinds of argument produced by headteachers and others in support of introducing mixed-ability classes clearly reveal their conviction that such a system promotes personal development or develops a socially cohesive community. They thus imply an acceptance of the view that teachers should be concerned not just with whether their pupils get 'O' or 'A' level passes, but also with what sort of people they are when they emerge from school, and they also display a conviction that this is best achieved by a mixed-ability form of grouping.

Lastly, we must note that this development is also in part a result of a change in our attitudes to values themselves. We have already referred to the fact that it suggests that we no longer see knowledge and human activities as hierarchically ordered. We cannot now be confident in claiming high status for any kind of knowledge or activity, another legacy of Platonism that we are belatedly shaking ourselves free of. The value of an activity can only be satisfactorily defined in terms of the value people actually place on it either as individuals or collectively as a society or a community. This must lead us to be increasingly hesitant to claim that the value we place on certain kinds of knowledge has any degree of objectivity or finality and, especially in a changing society, to be reluctant to be dogmatic in our desire to impose values on the next generation. Certainly we can no longer justify valuing people differentially according to the kinds of knowledge they possess or the kinds of activity they can cope with.

Mixed-ability grouping seems to me to imply an acceptance of this and a recognition of the need not to make one kind of educational provision for all based on some view we have of what is worthwhile and certainly not to make two or three offerings graded to suit two or three crudely defined types of intellectual ability, but rather to endeavour to offer to each child what he can come to recognize as being of importance and value to him. If that is what we want to do, there is no point in grouping pupils by ability; indeed, grouping by ability is likely to be positively counter-productive to these aims.

These, then, seem to me to be among the more significant features of the case for mixed-ability teaching. Firstly, it springs from a new view of men and society, of man as having the potential to become a cooperative social being and of society, as a result, as egalitarian in the broadest sense. Secondly, it is the result also of major changes in the view we take of what education itself is, changes which are in themselves, of course, interrelated with the changing ideology of society. Lastly, it is also a consequence of some major questions that are being asked about the nature of values themselves and the basis for making judgements of value about anything, especially about different kinds of knowledge, and a resultant desire to avoid dogmatism in the planning of a curriculum and to endeavour to make it meaningful to the individual child in the light of his own needs and interests.

If we are to produce a system of education that will fit the kind of ideology I have just described, which will promote the personal development of the individual, provide him with the ability to think productively for himself and society, encourage the growth of a more socially cohesive community and in every way meet the needs and the impact of a continuously evolving society, the keynote of that system must be flexibility. For it must both produce people who are themselves adaptable to change and it must itself be open to constant adjustment to the changes that are continuously taking place. Continuing technological and social change requires continuing educational change as well as the production of people well attuned to the realities of change. The system we are looking for, therefore, must do three things. It must be productive of adaptable citizens; it must in itself facilitate continuous development; and it must be responsive to the changing demands that are made on the school system. It is this kind of flexible system that is required, so that we must finally turn to an examination of the extent to which mixed-ability grouping can offer such flexibility, since only thus will it be possible to establish the right kind of positive rationale for such a system of grouping.

A flexible system of grouping

Again, we must note first the inadequacies of streaming when considered from this point of view. For one of its major characteristics is its total inflexibility. As we saw earlier, it is based on the strange concept of general ability and, as a result, it produces groupings which are used for all purposes and which are probably suitable for none.

Let us consider for a moment what happens when a large secondary school streams its pupils on entry. Selection for a suitable class will normally be based on the evidence of attainment, attitude and so on provided by the primary school of origin, taken in conjunction with that produced by any internal interviewing procedures the school itself uses. The objective evidence will usually take the form of scores on a mathematics test, an IQ test, an English test, and perhaps an essay. The results of such tests of attainment in these diverse areas are roughly balanced against each other; some consideration of other factors, such as motivation and attitude, gleaned from interview or from the assessment of the primary school head or teachers, may be thrown into the equation; and allocation is thus made to a group which is then normally used for all the multifarious purposes, both academic and social, of the school.

Thus there is produced a series of classes, whose rationale is intellectual and academic homogeneity, but which are far from homogeneous by any criterion. For such classes will not contain all the best mathematicians or English students even, let alone the best historians, geographers, scientists, linguists, artists or gymnasts. Yet they will be offered to every subject teacher as homogeneous groups carefully selected for his purposes. We know that few bright pupils are bright at everything and few slow learners are slow at everything. Yet these classes once fixed are usually used for all purposes, social as well as academic. Thus streaming offers us a system which is totally inflexible and there be no greater evidence of its inefficiency than the fact that some schools have found it necessary to 'set' children for certain subjects as well as streaming them in this general way.

This system we need to replace with one that offers a far greater flexibility of grouping, so that we do make it possible to produce the right groups for the right purposes. The purposes of a school are multifarious and many different kinds of grouping are needed to meet them, not only in terms of getting the right groups for the teaching of mathematics, English, French, science and so on, but also to create the right social groupings as well. If mixed-ability grouping, then, is to offer us the system we need, it must be shown to provide greater flexibility than we have hitherto enjoyed, so that the only way to produce a positive rationale for it is to show that it can offer this kind of basis for flexibility and for continuous change and development.

It is clear that no one form of grouping will ever be shown to be objectively better than any other, because no one form of grouping will be suitable for all purposes. The strongest argument for mixed-ability grouping, then, lies not in attempting to establish that it provides us with better teaching units for all purposes, especially as we have already referred to the fact that there are almost as many versions of mixed-ability groupings as there are schools practising this kind of organization. To suggest this is to advocate the establishment of a new orthodoxy, a new rigidity that in time must become as inhibiting as the old that it replaces. The strongest case lies in the fact that mixed-ability grouping presents us with a system that can and should be flexible enough to allow for the creation of different groupings for different purposes and to facilitate continuing development of all kinds.

We have seen some of the arguments for the social advantages of mixed-ability classes. These advantages will only accrue if these classes are working units, so that they must constitute the working groups for as many areas of the curriculum as is possible. Which areas these will be will depend and should depend on the local conditions that apply in any particular school and especially on the attitudes of its teachers. However, other groupings should also be possible and should be made available where they are felt to be necessary or appropriate. Most of these will be discussed in detail in later chapters. It will be enough, therefore, at this stage merely to note some of them. There is scope for constant regrouping within the class to meet changing needs and purposes; opportunities are presented by and for the withdrawal of pupils, either as individuals or in groups, for specific purposes such as remedial work of all kinds or extended work for the gifted; there are the advantages of setting, especially at the approach of public examinations; and there are the as yet largely untapped possibilities of team-teaching, which makes all kinds of groupings possible. The mixed-ability class is to be seen as a base upon which many other kinds of grouping can be built and it is this feature of it that constitutes one of its major justifications.

All of these devices offer the possibility of manifold adjustments of grouping to meet the manifold purposes and experiences of the school. They also offer scope for the adaptation of this system to meet the peculiar needs of each individual school situation. As we have seen, mixed-ability grouping means many things to many people and no definition that is universally applicable can be found. However, it is perhaps not desirable that we should achieve such a definition, since the effect of that might be to introduce another kind of rigidity. All that one can and should say is that our groupings need to be far more sophisticated than they have been hitherto and that they should include the possibility of constant adjustment to meet different needs.

The second aspect of the case that is being made out for mixed-ability grouping, then, is that it can provide a flexibility that makes possible constant adjustment in the light of the needs of individual pupils and individual schools as well as of society itself, that it enables the school to respond quickly and smoothly to the changing demands of society. This is the major advantage of this form of grouping and constitutes the main argument in the development of a positive rationale for such a system. In short, its chief merit is that it can assist continuous curriculum development by making this possible at the level of the individual school.

One of the reasons why school-based curriculum development is currently being advocated is that attempts at the dissemination of curriculum innovation by a centre-periphery model have achieved very little and have encountered many difficulties. If there is to be continuous curriculum development, therefore, to keep pace with the changes in society, it would appear that this process can only be a smooth one if it takes place within each school. Another way of putting this would be to say that this is the only way of ensuring that curriculum change is really curriculum development. As Malcolm Skilbeck says, school-based curriculum development 'provides more scope for the continuous adaptation of curriculum to individual pupil needs than do other forms of curriculum development' (Skilbeck 1976, pp. 93-94). Other systems are 'by their nature ill-fitted to respond to individual differences in either pupils or teachers. Yet these differences . . . are of crucial importance in learning. . . . Curriculum development related to individual differences must be a continuous process and it is only the school or school networks that can provide scope for this' (Skilbeck 1976, p. 94).

It is being argued further here that it is only a school that is organized on a proper mixed-ability based that can respond to these demands, because only such a school will offer the flexibility of structure that continuous adaptation of this kind requires. If we are convinced of the value of school-based curriculum development, then, we must be concerned to find the kind of organizational structure that will most facilitate it, so that we must look seriously at the claim that it will be promoted by the kind of flexibility we are suggesting a move to mixed-ability grouping should bring.

This, then is the kind of continuous adaptation it is being claimed society needs, so that, if mixed-ability groupings will help us to achieve it, we have here the strongest case for introducing them. We need a system that will produce people who are themselves adaptable to changing material and social conditions. We need a system that will facilitate that constant adjustment and curriculum development that a changing society and an evolving education system needs. We need a system that will not make change difficult or subject to violent conflict.

This is the kind of system we want and, whether we call it mixed-ability grouping or something else, it is only for this kind of system that a positive rationale can be produced.

Summary and conclusions

In this chapter I have tried to show that a rationale for mixed-ability grouping, especially in secondary schools, is now needed, that in part that rationale will be founded on the manifest inadequacies of streaming, that a more positive justification can be developed by reference to the changed and changing needs and ideology of both education and society and that the key factor in this case is that very openness to continuous change that should be a feature of any education system in an advanced technological society. I have further tried to show that, properly conceived and executed, this is the kind of system that can and should be developed from the introduction of what are currently being called mixed-ability classes. It is a pity that we have no better term at present for a system of grouping the main merit of which should be that it will not ossify or rigidify educational practice but will open the way to continuous modification in the light of social change and educational experience.

Most of the major implications of this for the practicalities of teaching will be examined in the chapters that follow.

CHAPTER 2

FIRST PRINCIPLES

Adequate practice in the classroom, as in any other sphere of human endeavour, must be based on a sound appreciation and understanding of relevant theoretical considerations. The precise nature of the relationship between the theory and the practice of education is a matter of considerable debate. Few would wish to dispute, however, that the simplistic model of education theory as a body of ideas or ideals to be assimilated and then somehow 'applied' in practical situations is not satisfactory in any way, not least because, since it does not work like that, most teachers who see the relationship in this way quickly come to the view that theory is unnecessary. Education theory has to be viewed rather as a body of theoretical understanding that gives us a wider perspective on practical issues — not only a broader range of strategies to use in the school or classroom but also a more sensitive and perceptive view of what we experience and observe. Understanding of this kind is necessary if the teacher is to be able to recognize the important things that are to be seen in the behaviour of his pupils, appreciate their significance and discern some of the reasons for them. For only from that kind of base will he be able to devise appropriate responses to them or methods of dealing with them.

Thus adequate practice in the teaching of mixed-ability classes will require some kind of understanding of the major theoretical considerations upon which this system of organization is based. For this reason, the attempt was made in Chapter 1 to produce a rationale for mixed-ability grouping by identifying some of the main reasons why many teachers and others have come to feel that this system offers a better device than others for achieving what they see as being the main concerns of education.

For the same reason, it is necessary now to try to pick out some of the general educational or pedagogical principles that this approach gives rise to — those fundamental principles upon which it is based. For if teachers do not understand these basic principles, they are unlikely to be able to ensure that their practice adequately reflects them. It is this that the present chapter will endeavour to do.

There are several things we need to do in order to achieve this. In the first place, if I was right in Chapter 1 to argue that changes both technological and social provide a major justification for the introduction of mixed-ability classes, we must now consider what they imply for the way in which we set about teaching pupils in these classes. Secondly, we must consider what is involved in the individualizing of educational provision that mixed-ability classes require. Thirdly, we must examine some of the implications of endeavouring by this method to attain more closely the ideal of equal educational opportunity for all. Lastly, since all of these considerations will involve major changes in our approach to teaching, we must look at some of the implications of these changes for the role of the teacher.

Social change and the mixed-ability class

Reference has already been made in Chapter 1 to the extent of recent technological change and to the fact that it leads in turn to social change of an equally drastic kind. Both of these have resulted in far-reaching changes in our life-style. What this means for the teacher is not only that he must prepare his pupils for a society very different from that in which he grew up himself, but, more importantly, that he must prepare them for change itself. In broad terms, it is the teacher's job to develop in children the skills that his society needs, whether they be hunting and fishing or precision tool engineering; it is also his task to hand on to them the values of the society and to initiate them into the traditional culture of the society and the knowledge upon which that culture is based. In all of these areas change in recent years has been so marked that teachers understandably are bewildered as to what skills, values and knowledge they are expected to hand on. If we develop in children the skills that present-day technology requires, we will find many of these are skills that will become obsolete within their working lives. What is needed, as we noted in Chapter 1 the Crowther Report (1959) stressed, is not a training in particular skills but the development of a general mechanical ability that will make adaptation to the changing requirements of industry a relatively easy matter, a flexibility that will be brought about only if teachers concentrate less on the inculcation of knowledge and more on the acquisition of understanding.

Similarly, in the realm of values there is no one code that the teacher can see it as his job to hand on, no fixed and final goal in this sense for moral education. We live in a pluralist society, a society in which there are many codes of values, and to prepare children for such a society is not to attempt to inculcate a fixed attitude to moral questions but rather to develop in them the ability to think morally and to reach their own conclusions on the ever new moral issues that a changing society will present them with. Again it is flexibility and adaptability that must be the aim and it can be achieved only by a concentration on the acquisition of understanding rather than of knowledge.

Finally, in the realm of knowledge itself, what is needed in an age of computers is not people who can store it, but people who can use it intelligently and contribute to its continued expansion. In all of these spheres it is clear that the teacher can no longer be merely a conservative agent for the transmission of traditional skills, values and knowledge; he must set about providing children with the abilities that will ensure continued development on all of these fronts. He must look to the future rather than to the past.

It is equally true that social change has brought with it a need for teachers to be more aware of the social and emotional development of their pupils rather than merely concentrating on their intellectual development. Again we noted in Chapter 1 that this is one of the reasons why some people have felt that a change from a streamed to an unstreamed form of organization is desirable. However, it must be stressed that developments of this kind imply no reduction in the importance of the teacher's role as instructor. There is no reason to assume that less emphasis needs to be placed on learning or on intellectual achievement in a situation where the aim is to attend also to pupils' social needs and where the mixed-ability group is the basic teaching unit. Sociologists have distinguished between the 'expressive' and 'instrumental' functions of the school, between its social and pastoral responsibilities and its duty to promote study (Musgrove 1966). Clearly, this distinction is important, but it is often too readily assumed that this can be the basis for distinguishing the primary from the secondary school, the secondary modern from the grammar school, the education of 'working-class' pupils from that of 'middle-class' pupils and the ideals of the college of education from those of the university department of education. Too readily it is assumed that the main objective of the primary and of the secondary modern school is the social welfare of its pupils and that their intellectual advancement is of secondary importance; too often it is assumed that colleges of education, preparing teachers mainly for primary and nonselective secondary schools, will stress the social and pastoral role of the teacher to the detriment of his duties as an instructor; too frequently it is assumed that a teacher who evinces a concern for the social welfare of his pupils is *ipso facto* ignoring their intellectual needs.

As a result of this rather naïve dichotomy, there is a tendency to assume that the main purpose of a teacher of a mixed-ability group should be social, pastoral, 'expressive'. There is no justification whatsoever for this assumption. The advent of the mixed-ability class will, of course, make it necessary for him to take a broader view of his duties and recognize the need to include such considerations in making provision for his pupils. He must begin to look as the curriculum not only from the point of view of certain logical requirements of the subject-matter, but also in the light of the psychological and social requirements of his pupils and must accept responsibility for their moral, social and emotional as well as their intellectual development. If he does this, he will soon realize that it is not possible in practice to distinguish these, since there is an interrelationship between learning and feeling, between the cognitive and the affective, that cannot be ignored if he is to attend successfully to either need. However, there is no reason why the introduction of a mixed-ability organization to a school should bring with it any reduction in intellectual rigour or in the academic demands to be made by teachers of their pupils. The teacher's prime job is to teach and this is not altered by this kind of organizational change, whose purpose rather is to enable him to teach more effectively. It is *how* he is to teach that is changed and we must turn our attention to what is implied for that.

Individualized education

All the implications for changes in the approach to teaching required in a mixed-ability class derive from the fact that such a class is by definition heterogeneous. It is a well-known, although usually disregarded, fact that all classes are heterogeneous, even in situations where the streaming is very fine. Nevertheless, for economy's sake, teachers have tended to ignore this fact, to regard streamed classes as homogeneous and to teach their classes as units, 'aiming at the middle' and dealing with the 'top' and 'bottom' as best they can. No teacher faced with a mixed-ability class can hope to keep up this kind of approach for long. 'Top' and 'bottom' will be too far apart. It will be necessary, therefore, for him to view his class as a collection of individuals and to recognize individual differences of all kinds.

Although psychologists have been stressing the importance of individual differences for education for forty years or more, they have tended to be largely concerned only with differences of intellectual ability and teachers in practice have tended to follow suit. Differences in rates of progress and in amounts of material that can readily be assimilated have been taken fully into account by all concerned with education — this after all is the justification for streaming and selection; it has been accepted that children will progress at different speeds and that for some the work will need to be watered down, some-times to a very thin gruel; but no other differences have been regarded as educationally significant. If they are not significant, there can be little justification for doing away with streaming, since streaming is designed to enable teachers to cope most efficiently with different speeds of learning and different points of cut-off in a situation where all children are to learn the same and to proceed along the one educational highway as far as their intellectual fuel will carry them.

To make away with grading of this kind implies a wish to take account of more subtle differences between individual children and a conviction that the content of education must be varied to suit individual needs. In other words, to abandon streaming, to embrace from choice the clearly heterogeneous teaching group is to make a commitment to the notion of individualized or personalized learning, to providing for each child according to his or her unique educational requirements. This is what many have meant by the expression 'child-centred' education. 'Learner-centred', 'individual-centred' or person-centred' might be better terms, since too often 'child-centred' has rather naïvely been taken to imply a 'mothering' approach and little more. What many 'child-centred' theorists have been advocating for a long time is that educational provision should be adapted to the individual pupil. In the mixed-ability class this becomes essential and its implications need to be looked at in some detail.

In the first place, there are quite serious implications here for the content of education. There was a time when all discussions of curriculum centred on decisions concerning its content. This is why such discussions did not get very far; they were fundamentally mistaken in that they saw the curriculum only in terms of its content and this concern with content blinded them to questions concerning the goals of education or the educational principles upon which particular decisions should be based. To plan a curriculum, then, was to set out a programme of content, and inevitably such curricula were planned for groups rather than for individuals.

Recent work in the field of curriculum study has highlighted questions of goals and principles and has rightly seen these questions as being central to curriculum planning and logically prior to questions of content or method. Decisions about the content and methods of education can only be taken when we have achieved a clear view of our educational goals or the principles that are to inform all our practices (Hirst and Peters 1970). Whatever the source of these goals and principles, whether they are socially determined or derived in some way from an analysis of the notion of education itself, we must be clear about them before we can consider any of the other questions that face us as curriculum planners.

We must recognize, of course, that in practice there will always be a significant difference between our intentions and our actual achievements, not least because our pupils' intentions and those of their parents will differ from our own, and that our own purposes will be modified by our practical experiences; but curriculum planning is essential if education is to be a purposeful activity and it is goals and principles that constitute both the central and the common element in such planning.

Once we concentrate our attention on them, we can see the secondary status of content, we can see that what is important is not that we should strive to ensure that all children assimilate a certain body of knowledge, but that we should be endeavouring to lead them to the same educational goals; and there may be as many roads to these goals as there are individual children. What matters is not that all children should have a knowledge of the main events of the French Revolution but that they should all be able to think historically and have a developing understanding of the problems that beset man's attempts to learn to live with his fellows; not that they should all 'know' Boyle's Law, but that they should all be able to think scientifically and understand how man's knowledge of his physical environment has been and is being developed; not that they should all have learnt Hamlet's soliloquy, but that they should have been introduced to and brought to value literature which is capable of interpreting and enriching human life. As Jerome Bruner has said, 'We teach a subject not to produce little living libraries on that subject, but rather to get a student to think mathematically for himself, to consider matters as an historian does, to take part in the process of knowledge-getting. Knowing is a process, not a product' (Bruner 1966, p. 72).

However, if we accept that the goals and principles of education are the same for all children, it is not necessary to assume that the *content* must be the same for all. This has been the mistake of the past. We have attended only to intellectual differences in our pupils, as we have seen, and have been prepared to modify only the rate of progress we expected of them or the amount of content we expected them to digest. In doing so, we have pretended we were individualizing our approach — the growing use of teaching machines is welcomed by some as an individualizing of learning in this very limited sense — but this is far too unsophisticated and is, in fact, a largely illusory form of individualization. If we accept that only the goals of education are common, we must be prepared for quite dramatic differences of content and method to suit the dramatic differences that clearly exist in children's styles of learning, levels of motivation, interests, backgrounds, ambitions and the many other facets of their unique personalities.

Many teachers have, of course, tried very hard to adapt their approach to the needs of individual pupils, but the system has too often made difficulties for them. The advent of the mixed-ability group should have helped to remove some of these difficulties, but it can only succeed if teachers themselves can shake off some of their preconceptions concerning the content of education and move towards a greater clarity of thinking about the curriculum. Once they have freed themselves of the limitations imposed by this narrow view of content, they can take advantage of the greater scope for manoeuvre that a mixed-ability group provides and the greater opportunities to adapt both content and method to the individual.

A second point about content arises from what has just been said. Just as we can no longer justify a belief that the content of education should be the same for all pupils, so we can no longer assume a unique sequence of learning even for the same material. There are no grounds for the assumption that all knowledge is organized hierarchically, that children must be taken through it step by step or, to change the metaphor, that they must build their knowledge brick by brick. This is another example of that folklore that has for too long hindered the establishment of a serious theoretical basis for the practice of education. Too readily it has been assumed that children cannot learn 'y' until they have learnt 'x', or that certain 'subjects' cannot be broached before certain others have been begun or that certain kinds of topic cannot be tackled before a certain age has been reached.

To some extent the work of psychologists such as Piaget on the development of children's thinking has reinforced this kind of view, but perhaps their findings have been interpreted with too little sophistication. Again there is a confusion of ends and means. Their work would certainly seem to suggest that there is a definite sequence to the acquisition by children of certain kinds of concept, and therefore that there are certain psychological considerations to be taken into account in curriculum planning. It does not follow that there exist psychological factors which rigidly control the sequence in which we must present pupils with the content of their education. A great deal more work needs to be done in this area, but in the meantime it would seem safer to assume with Jerome Bruner (1962), that 'any subject can be taught to anybody at any age in some form that is honest and interesting' (1962, p. 124); that it *is* possible to devise versions of any subject that will make it suitable for presentation to pupils at any age or stage of development, if for other reasons it seems warranted. The same material can and should be presented again later in a different way so that gradually the conceptual progress we are after can be attended to and our educational goals achieved. This process of covering the same ground again and again at different levels of complexity Bruner calls the 'spiral curriculum' (1960, 1966). It is a notion that is worth a great deal of careful attention.

This consideration of the work of the developmental psychologists in itself also adds to the case not only for mixed-ability grouping but also for an individualized approach to teaching. We saw in Chapter 1 that one of the criticisms that has been made of streaming is that it is based on a psychometric view of intelligence, a view that has been shown to be at the very least questionable by the work of Piaget and others on cognitive growth. We must now also note that their work has revealed that there is no close correlation between children's chronological age and the age at which they reach the various stages of cognitive development. To try to teach groups of children of any one chronological age as homogeneous units, however they are selected for grouping, would, therefore, seem to be misguided. Consideration of their needs as individuals would seem to be essential if only in terms of the variation in their rates of development.

A final point to be made here about the implications of these changes in approach for the content of education is that it would seem that in such a situation traditional subject barriers must lose some of their strength. Again this might appear *prima facie* to be flying in the face of logic, denying the validity of the logical differences that clearly exist between the disciplines. Again, to a large extent, all we are flying in the face of is folklore and again it is a folklore that has confused ends and means. Our intention may be to get pupils on the inside of certain distinct forms of knowledge, but there is no evidence to suggest that the only or even the best means of achieving this is by organizing the curriculum into separate subject areas.

It has been argued that the best way to attain this goal and at the same time to ensure that links between subjects are not ignored is to establish a curriculum which is 'fundamentally subject-based, but which pursues links between the different disciplines with real seriousness' (Hirst 1967, p. 83). It is being argued here that the logic of the mixed-ability class makes it necessary to approach the problem from the other end, to ensure that pupils are made aware of logical differences between disciplines as they pursue their individual programmes, in such a way that they see the point of these differences and recognize their significance in the organization of their own knowledge.

Once again much work needs to be done in this area, but once again we would be well advised not to prejudice the issue. Our main concern as teachers is to discover the most efficient means of educating individual pupils and we must not without good reason let ourselves be restricted in our decisions by other considerations.

Let us now turn to a consideration of the implications of the mixed-ability teaching situation for teaching method. Clearly, what we have said about individualizing content will apply to method; methods, like content, should be adapted to the learning styles of individual pupils and we will look in detail later at the practical implications of this. However, there is one general point about method that needs to be made at this stage since as a general principle it will inform all our practice. It will be a basic rule of our approach to learning and teaching in this kind of situation that we must lean towards heuristic rather than didactic methods. This point is made here not from ideological considerations nor as logically entailed by positions we have already accepted but as a point of sheer practical necessity. If we accept the view that every pupil must be helped to advance at his own pace by his own route to the goals of education, then, unless we assume a much more favourable pupil-teacher ratio than is ever possible or even, perhaps, desirable, we must conclude that there are real limitations on the time a teacher can spend 'teaching' any one pupil and that of practical necessity each must learn to work on his own.

It is in my view a mistake to make a great thing of 'discovery methods', to talk rather loosely about 'learning to learn' and so on, not least because it creates another dichotomy to add to the many that litter educational theory, most of which arise from a lack of subtlety in our thinking. We ought not to be asking whether instruction or discovery-based methods are more appropriate in education and thus generating an artificial conflict between the two; our energies ought to be directed towards discovering what combinations of the two approaches are most effective and under what conditions. The education of a person is neither an imposition on him from without nor a natural development from within; it is a combination of the two as the growing, developing individual reacts to his environment. As teachers, therefore, we ought not to be torn asunder by asking ourselves whether we should 'teach' our pupils or let them 'find out' things for themselves. It is impossible to learn anything without both of these elements. We ought to be experimenting with both approaches, remembering that the most effective combination for each pupil will depend on his own style of learning. Is it best to give him a few facts and some clear indications of how he can go beyond them on his own or does he need to be fed a lot of straight factual information? These are the questions teachers should be asking themselves as they plan the work of individual pupils. There will still be many occasions, even in mixed-ability classes, when some straight class teaching will be appropriate, but often in this kind of situation 'teaching' will mean organizing the learning of individual pupils, and they must of necessity work for much of the time on their own or in small subgroups.

Nor is it neccessary to assume that we should organize our programme in two parts, one devoted to the pursuit of interests and individual projects, the other to the acquisition of 'basic skills', as is done in many primary schools. Such skills cannot be taught in isolation at anything other than a basic level; indeed, if they are to be seen as a part of education in the full sense, if reading, for example, is to be not just a mechanical performance but is to involve comprehension, understanding and, indeed, a love of reading, the learning of the skill cannot be isolated from other aspects of the activity of which it is a part (Blenkin 1977). Thus teachers must be aware of the need to develop these skills through the interests and individual projects of pupils in much the way that many infant teachers set about the teaching of reading, writing and number.

A corollary of this insistence on individualizing education is that the pupil is being asked to take an increased and ever-increasing responsibility for his own learning. This is as it should be, since if we are concerned to educate, then part of what it means to be educated is to be autonomous, to be self-propelled. If our pupils learn everything under duress and compulsion, they may in the end be very knowledgeable but they will never be educated. Furthermore, as we have seen, the changing nature of society makes autonomy a vital economic aim of education, since a changing society needs autonomous citizens. Thus practical necessity is here the ally of our educational ideals. If education is our concern, then one of our goals must be the autonomy of the learner. From the beginning we must try to create a situation in which our pupils' learning is increasingly self-directed and self-propelled, until eventually that education can go on without us; we become superfluous. It is the educator's job to render his own services unnecessary.

There are, of course, a number of difficulties inherent in this approach to education. To begin with, it will be apparent that pupils from 'working-class' backgrounds are unlikely to take naturally to this method of learning. A facility with language is required for learning of this kind that many recent studies have suggested the 'working-class' child does not have. A pupil whose conversation at home is confined within a restricted code of language is bound to find it more difficult to engage in self-propelled learning (Bernstein 1961). Some kind of elaborated code may be acquired through education, but it will perhaps never be 'natural' and its acquisition will require a great deal of direct personal contact with a teacher. The social climate of those homes with a tendency towards somewhat authoritarian and arbitrary relationships also is likely to breed dependence rather than independence of thought within the child, and therefore will not promote the child's ability to take responsibility for his own work in school (Klein 1965).

How much easier is it for pupils from 'middle-class' homes? Certainly it must be somewhat easier, since many of the disadvantages just mentioned will be absent, but it can never be entirely easy for anyone. The solution, however, cannot be the rejection of this approach. We cannot reject it since, as we have seen, the very logic of education forces us to accept it and so much of the work of educationists in recent years has revealed it as the only basis for learning. What we need to do is to accept self-direction as an aim but to realize that with all pupils its achievement will be a slow process and with some a very slow process indeed. Much of the pupil's work, especially in the early stages, will need to be teacher-directed, but the aim throughout will be eventual self-direction. Again much more work must be done to help teachers with the practicalities of leading their pupils towards this goal. Some of the practicalities we will look at in later chapters. For the present it is sufficient that we make the goal explicit and stress what was stressed earlier, that no loss of rigour is entailed. If the self-directed learning degenerates into self-directed playing or messing around, if the rigour and the learning disappear, then we are as far from achieving our goals as we are when we take a highly didactic stance from the outset.

Equality of educational opportunity and the mixed-ability class

This reference to some of the problems faced by children from 'working-class' backgrounds takes us naturally on to the next aspect of mixed-ability grouping we planned to examine, namely its implications for the achievement of something nearer equality of educational opportunity than has been achieved in the past.

It will be clear from what was said in Chapter 1 that this is a basic feature of the rationale of mixed-ability grouping. The major objections to streaming which we considered there were directed at those features of that system that have resulted in an unequal distribution of educational opportunity and provision, while, conversely, one of the main elements in the case that was put there for mixed-ability teaching was the evidence that this does enable a higher quality of education to be made available to a wider range of pupils. This must, therefore, be one of our first principles in teaching mixed-ability classes.

We have already considered one aspect of this in looking at the problems of selection. Clearly, equality of educational opportunity is not forwarded by a system that is based on the use of selective procedures that are riddled with inaccuracies and which result in the misallocation of large numbers of pupils. The abolition of selection, then, both for different kinds of school and for different kinds of class within those schools should represent a major step towards the achievement of equality of opportunity in education.

There is another aspect of this issue, however, that we must develop here because it has crucial implications for the way in which we set about organizing the teaching of mixed-ability classes. As we noted in Chapter 1, one of the reasons why the misallocation of pupils within a selective system has such widespread effects is because of the generation of different curricula for different schools or for different streams. This has several aspects. Firstly, it makes the initial selection much more decisive than it need be, since it is a major factor in that absence of transfer that was also discussed in Chapter 1. If allocation to a particular school or stream means that you begin to work on a different curriculum, then clearly subsequent adjustment through transfer becomes very difficult if not totally impossible. Secondly, it represents a tacit admission that there are large categories of pupil for whom education in the full sense is not possible. It seems to accept that, as in Plato's ideal state, only the children of gold can be educated fully. Thirdly, because of this, it results in the generation by the educational system itself of social divisions, based on a stratification of knowledge (Young 1971) and the emergence of two or more cultures, a process which is accepted by some and even advocated as desirable (Bantock 1968, 1971) and one which, although opposed by others, is nevertheless recognized as an important by-product of streaming (Keddie 1971; Esland 1971).

The effects of this process of developing separate curricula have not always been fully appreciated, but that this has been the dominant feature of curriculum planning at least at secondary level in the United Kingdom is clear from the briefest survey of the recent history of curriculum development. Traditionally, the curriculum of the elementary school was quite different from that of the public and grammar schools, and it is quite plain that the advent of secondary education for all was not intended to change this. For those reports which recommended a tripartite system of secondary schooling had very clearly in mind the development of quite different curricula (Hadow 1926; Spens 1938; Norwood 1942). Indeed, among the early criticisms of the secondary modern schools, predominant was the charge that they were 'aping the grammar schools' rather than developing their own distinctive programmes, and it was because they were expected to be doing something quite different that for many years they were discouraged from entering their pupils for public examinations.

Then in 1961 the Central Advisory Council for Education (England) was asked to prepare a report on the education of pupils aged thirteen to sixteen of average and less than average ability, and under the chairmanship of Sir John Newsom produced in 1963 a report on *Half Our Future*, a title which indicated clearly enough in itself that this half was seen as requiring very different provision from the other half. The general theme of this report was that such pupils should be exposed to courses which are relevant and realistic, often with a largely vocational bias and with an emphasis on personal and social development. There should be resistance to external pressures 'to extend public examinations to pupils for whom they are inappropriate, or over an excessively large part of the programme for any pupil' (p. xvii). Thus the report has been criticized for accepting the principle of allowing the generation of a separate curriculum for the supposedly 'less able', for thus accepting a differentiation of the curriculum and a stratification of knowledge based largely on social grounds and for appearing to recommend that the curriculum for these pupils should emphasize conformity to their status in life and an acceptance of the norms and values implicit in it.

The work of the Schools Council has tended in the same direction. Established in 1964, it inevitably saw its role as being largely to assist with the implementation of many of the Newsom proposals, particularly those related to the raising of the school leaving age to 16. Some of its pronouncements on the curriculum have expressed the ideal of a set of curriculum proposals that could be applied to the education of all pupils. For example, its Working Paper No. 2, *Raising the school leaving age* (Schools Council 1965) speaks of 'the possibility of helping pupils who are the concern of this paper to enter the world of ideas, to use powers of reason, and to acquire even the beginnings of mature judgement' (p. 9). However, it goes on to suggest that this ambition 'may seem to contradict the experience of many teachers' and, indeed, the practice of schools had already hardened after the advent of the Newsom Report and most of them were busily developing 'Newsom courses' for those whom they sometimes even called their 'Newsom pupils'.

Thus the Schools Council's follow-up paper to Working Paper No. 2, Working Paper No. 11, *Society and the young school leaver* (1967a), offered some examples of what it felt were good courses for the 'less able' pupil, among which the project on 'The 97 Bus' attracted most attention as epitomizing all that many people have found reprehensible in this kind of approach. For it seems to suggest that we should offer in the name of curriculum development to those children we think we can label as 'less able' low status knowledge of a kind which, like the traditional curriculum of the elementary school does not encourage them to think critically for themselves but rather to accept their position in life and the social status and environment to which they were born (White 1968); in short, that we should not try to provide them with an education in the full sense at all.

This has been true not only of the Schools Council's attempts to offer advice on the development of a suitable curriculum for the young school leaver; it has also been true in a general sense of its whole approach to curriculum development. For most of the projects it has sponsored have been directed towards the 'less able' pupil. Clearly, there are other considerations, such as the ease of winning acceptance for new proposals, which have contributed to this, but its effect has been to emphasize this kind of differentiation of knowledge.

Again, there would seem to be two aspects of this. The introduction of new areas to the curriculum has been often limited to the work of the 'less able' pupil, so that projects in moral education, social education, integrated studies and so on have been set up and their approach geared largely to the needs of such pupils and on the assumption that the 'more able' do not need this kind of education. Secondly, projects that have been developed in traditional subject areas have often been aimed at the same group of pupils, as was the case with the Geography for the Young School Leaver project. Thus the tendency has been towards the generation of two curricula divided from each other both in terms of the subjects they contain and, where the subjects are the same, the level of attainment and the kind of approach and content to be adopted in the study of these subjects. This process has, of course, been reinforced by the establishment of two levels of public examination with the advent of the CSE for those pupils not deemed capable of taking GCE 'O' level.

It could be argued that this has been to the disadvantage of pupils at both ends of the ability spectrum, since it has resulted in all pupils receiving less than that which it would seem they should be being given in the name of education. 'The lower ability groups have suffered from social adjustment courses, low status knowledge which is unexamined and therefore restricts occupational choice; whilst most aspiring to higher education would benefit from the erosion of some subject boundaries, the introduction of more active methods of learning and the use of relevance as one criterion for selecting content (Schofield 1977, p. 36).

More recently, however, as the implications of this approach have been drawn to our attention with growing force, there has been a reaction against it. The work of such people as those concerned with the Humanities Curriculum project has suggested that it is not idle to look for the development of powers of autonomous thinking in all pupils. Those concerned with the Geography for the Young School Leaver project have seen an 'O'-level examination established by the Southern Examinations Board which has thus made that approach to the teaching of geography and the materials developed in conjunction with it available to pupils of all abilities. The explorations of the possibility of a common examination at 16+ (Schools Council 1971) have reflected this same line of thinking. Finally, among the arguments currently being put forward in support of the idea of a common curriculum are some based on the points we have discussed here, those that are directed at avoiding the social divisions and the inequalities that are implicit in the generation of curricula that are based on this kind of stratification of knowledge and the division of it into low- and high-status subjects (White 1968, 1973). This case is now, of course, gaining impetus from the support of the economic and more overtly political arguments for the establishment of a common core to the curriculum that have emerged from recent public debates at national level on this issue.

All of these developments must be seen in parallel with the advent of comprehensive education and mixed-ability classes. For this is the fundamental rationale of both of those innovations. This is one reason why it is argued that the introduction of comprehensive education requires not only common schools but also common classes. For the principle of both is that every pupil should be given access to education in the full sense of the word and offered as many educational advantages as he or she is desirous or capable of profiting from. The move to mixed-ability classes then must be seen as a major contribution towards the move away from the stratification of knowledge with its resultant social divisions and towards the provision of a greater measure of equality of educational opportunity for all. This must be regarded, therefore, as another major first principle of teaching mixed-ability classes.

The implications of this for the practice of teaching such classes are far-reaching, as we shall see in subsequent chapters. We must note some of them here, however. For it is clearly important to recognize at the outset that if the move to mixed-ability classes represents an attempt to get away from the generation of different curricula, however defined, it will be ineffective if, having abandoned selection by ability in allocating pupils to schools and to classes within those schools, individual teachers then allocate them to different groups within the class on the basis of their perceived abilities and proceed to generate different programmes for them at that level.

Secondly, and conversely, it is not necessary to interpret this as a claim that all should have the same educational diet. The attainment of equality of educational opportunity does not require identical provision for all pupils; this would be a very naïve and unsophisticated interpretation of that notion (Downey and Kelly 1975). Similarly, we must be careful that we do not allow these arguments against the generation of differentiated curricula to cause us too readily to accept the very different arguments that are currently being put forward for political reasons for a common core to the curriculum defined in terms of curriculum content. We must be clear that what we are led to by the considerations we have just discussed is a reiteration of what we said earlier, that all pupils have a right to education in the full sense but that the route to that goal may be very different for each individual pupil. Thus we should be led to see the idea of a common curriculum in terms of its goals and the principles underlying it rather than in terms of providing a common diet for all (Kelly 1977).

Again, some of the detailed implications of that will emerge in subsequent chapters. We must merely note it here as one of the most important first principles of mixed-ability teaching.

Mixed-ability classes and the role of the teacher

Let us finally turn to a consideration of the implications of these changes for the role of the teacher and the conditions under which he does his job. For many reasons, both social and historical, the teacher in the British-maintained school has tended, consciously or not, to see his job as being concerned mainly with the promotion of the able pupil rather than encouraging the progress of all his charges. Historically, education in Britain has been a privilege granted to an increasingly large proportion of the populace (in contrast to the United States, for example, where it has for long been every child's democratic right upheld by the community), and British society has long been dominated by elitism rather than egalitarianism and an emphasis on what sociologists call the ascription of roles rather than their achievement (again in contrast to the United States where it has long been believed that status should not be ascribed by some criterion such as birth but should be achieved by individual ability) (Hoyle 1969). For a long time, therefore, many British teachers have tended to emphasize their selective and elitist role, the teacher in the primary school, for example, concentrating on the allocation of children to different types of secondary school, the secondary-school teacher concentrating on passing pupils on to various forms of higher education or into various types of occupation.

The move towards mixed-ability grouping implies a move towards equality and the achievement of roles, and the teacher's role must change accordingly. In a situation where learning is being individualized, where all pupils are not pursuing the same work in competition with each other but pursuing their own paths to learning, the teacher must renounce his selective, elitist role and accept responsibility for the educational advance of all his pupils on a broad front. This is perhaps the greatest professional reorientation that is required of him by these changes.

Secondly, an increased responsibility falls on him for decisions about the content of education. The teaching profession in Britain has always enjoyed a high level of autonomy in the matter of curriculum content and, because of the nature of his work, the individual member of the profession has also had a high degree of autonomy in this field. Inspectors, advisers, the Schools Council and other bodies can, and do, offer advice, comment and suggestion, but in the end it is the headteachers, the heads of departments and the assistant teachers themselves who decide what the content of the curriculum will be. When the emphasis in curriculum planning is on content, they can do much of this collectively, but when content is individualized, as we have suggested it must be in the mixed-ability group, a great deal more responsibility devolves on the individual teacher. He must be prepared to make many more decisions concerning what is appropriate for individual pupils in the light of their unique needs and the overriding goals and principles of education. Like his pupils, he himself must become far more self-propelled.

This will place a far greater premium on his understanding of the theoretical bases of education than may hitherto have been necessary. It has already been suggested that the model of educational theory as a body of ideas or ideals to be in some way 'applied' in the classroom is a wrong one. Rather it should be seen as offering the teacher deeper perspectives on his work that enable him the better to recognize its implications and to reach his own decisions. It will be clear that in a teaching context such as is being described here this will be far more important and necessary than in one in which he is teaching to a syllabus drawn up elsewhere. It is, of course, precisely for this reason that we need to discuss the basic principles of mixed-ability teaching before we can get down to its practicalities.

Thirdly, it will probably be apparent that in a classroom where learning of the kind we are describing is going on, there will be far greater informality than in a classroom where the teacher's main offering is the formal lesson. Where each child is doing his own thing, there must be a relatively informal atmosphere. This creates a situation which is far more demanding for the teacher. Apart from the demands that it makes on his intellectual resources and methodological expertise, it is a situation in which a completely different pattern of authority exists. The teacher must be *an* authority rather than rely on the fact (sometimes a somewhat doubtful fact) that he is *in* authority; his authority must be earned since it cannot be given to him and to earn it in this kind of situation is no easy task. There is already some evidence to suggest that it places a much higher premium on the teacher's personal qualities (Musgrove and Taylor 1969). It certainly makes heavy demands on his skill as a teacher and on the relationships he is able to develop with his pupils.

These points are well summed up by Basil Bernstein (1967).

There is a shift — from a pedagogy which, for the majority of secondary pupils, was concerned with the learning of standard operations tied to specific contexts — to a pedagogy which emphasises the exploration of principles. From schools which emphasise the teacher as a solution-giver to schools which emphasise the teacher as a problem-poser or creator. Such a change in pedagogy (itself perhaps a response to changed concepts of skill in industry) alters the authority relationships between teacher and taught, and possibly changes the nature of the authority inherent in the subject. The pedagogy now emphasises the *means* whereby knowledge is created and principles established, in a context of self-discovery by the pupils. The act of learning itself celebrates choice.

Finally, we must note a further point that Basil Bernstein is also concerned to stress in that same article. This shift in pedagogy reflects or is paralleled by a similar shift of emphasis in the principles and form of social integration in society as a whole. That shift is from what Durkheim called mechanical solidarity, which emphasizes such things as a common system of beliefs, detailed regulation of conduct and social roles which are assigned or ascribed, to what he called organic solidarity, which emphasizes such things as differences between individuals, a pluralism of value systems and social roles which are achieved. In short, the shift is from the 'closed' school to the 'open' school and this reflects a similar change in the bases of the society in which the schools are to be found.

This brings us back to the point with which we began our discussion of the first principles of mixed-ability teaching. We are now at a deeper level and can recognize some of the deeper implications of what we said then about the changing nature, needs and demands of society.

The introduction of mixed-ability classes represents both a reflection of what is going on in society and an attempt to respond to it. If we are to take full advantage of the opportunities it offers and ensure that it fulfils the purposes for which it is being established we must understand in full its implications.

Summary and conclusions

In an attempt to help teachers gain this understanding of the full implications of mixed-ability teaching this chapter has tried to identify some of its first principles. In doing so, it examined first its links with social change, both particular changes and the nature of social change itself, and it was suggested that this required a complete rethinking of the curriculum and of what the teacher must see as his responsibilities. We then considered some of the implications for this of the need to individualize educational provision, recognizing that this had to be seen in relation to the content as well as the method of education and that this consideration highlighted the need for us to be clear about our educational goals and those basic principles that would underpin our educational practices. Thirdly, we looked at the implications of the fact that the introduction of mixed-ability classes was to be seen, along with the establishment of comprehensive education, as a further device for the achievement of something nearer to the ideal of equality of educational opportunity for all pupils and, in doing so, we noted the need to view the education of all pupils in the light of the same educational principles rather than to see it in terms of the generation of different curricula to meet the needs of different 'types of mind'. Lastly, we considered some of the changes in the role of the teacher that were the inevitable consequences of these major changes in our approach to education and noted that they represented changes in the social relationships within the school which in turn reflect similar changes in the wider context of society itself, a point that brought us full circle to the place from which we had started.

In this way I have sought to identify some of the principles underlying mixed-ability teaching to provide teachers with that theoretical understanding that is essential for adequate practice. We may now have, therefore, a base from which we can make excursions into some of the more detailed practical issues that the teaching of mixed-ability classes raises. It is to these that attention will be turned in subsequent chapters.

CHAPTER 3

INDIVIDUAL AND GROUP ASSIGNMENTS

It is sometimes said that mixed-ability grouping requires teachers to appreciate and understand the differences between the children in their classes and to know their pupils as individuals. In the eyes of some, there can be no greater condemnation of streaming than that those who make such claims can assume that such an approach is not necessary in dealing with a streamed class. For there are many who would want to claim on both theoretical and practical grounds that all true education is individual education, that what is valuable to the individual in what the school offers him or her is that which is, or in some way becomes, personalized, and that, as a result, in any educational situation, teachers should be aware of the needs of each individual pupil.

Whether one accepts this or not, however, it is impossible to deny that the sheer practicalities of a mixed-ability class at any age or stage of education will necessitate that a good deal of the work be tackled at an individual or group level. There must be severe limitations on the amount of work that can profitably be undertaken at the same level by the whole class. The technique of working through individual and group assignments, therefore, becomes a very important weapon in the armoury of the teacher of a mixed-ability class.

Panaceas, however, are as difficult to find in education as they are in any other sphere and there is no one answer that can be given to the question, 'How do I set about teaching my mixed-ability class?' Every child differs in important respects from every other, so does every class, every school, every neighbourhood, as well as every teacher and head-teacher, so that the approach to education must be adapted to the particular conditions that apply in any one situation at any one time. Every school and every teacher will and should develop an individual style and deal with a unique situation in a way peculiarly appropriate to it. It is the sameness of so much secondary education that makes one wonder how it can be satisfactory for so many different pupils.

In any attempt to present an overall view of what is entailed by an individualized approach to education, therefore, there is the ever-present danger of encouraging the growth of a new orthodoxy, a new gospel, which may soon become just as restrictive as the old and may prove just as effective in limiting the innovatory role of individual teachers or groups of teachers. This must be avoided at all costs, since the education of all our children can only be forwarded by placing more responsibility for it in the hands of the teachers who are in personal contact with them. We have suffered for too long from education by remote-control.

It is, however, possible to offer teachers some guidelines as to how they might set about this kind of teaching, to offer them some tools which may enable them to do their job to their own greater satisfaction. To provide a carpenter with a chisel is not to tell him how to use it nor in any way to limit his creativity; rather it is to widen his scope and extend the range of his opportunities. It is in that spirit that the practical suggestions which follow are offered.

A further preliminary point is that these suggestions are offered with the aim not of replacing existing techniques but rather of supplementing them. For example, it is sometimes felt that once streaming is abandoned class teaching must also cease, that the teacher should at no stage be seen teaching his class as a whole. This again springs from too unsophisticated a view of the teacher's task. There is no evidence to suggest that class teaching is always ineffective when used with mixed-ability classes. Many teachers still make use of it most effectively. Some indeed are concerned about the implications of group or individual methods, since these can lead to the accentuation of individual differences and the aggravation of precisely those social disadvantages that unstreaming is concerned to avoid (Worthington 1971). Class teaching then, still has its part to play in the teaching of mixed-ability groups; indeed, there may be occasions when a lecture or other such presentation to a whole year group of pupils will be appropriate. If the emphasis in what follows is placed on group and individual teaching methods, it is because these are relatively new to many teachers so that it is here that practical advice is needed. They are suggested, however, as additional teaching devices that will add flexibility to the teacher's approach and not as new methods advocated to oust the old and tried.

Finally, it must be stressed that what is discussed here should be seen as having relevance and point for the work of any teacher in any class-room. Individual and group methods can be used to effect with streamed as well as with unstreamed classes, within subjects as well as in general teaching or interdisciplinary situations, in schools where the general approach is formal as well as in those where new methods are being tried. Clearly, the teacher who wishes to work in this way will find it easier to do so in a school where the organizational structure is designed to promote this kind of approach, where, for example, the timetable provides him with the substantial blocks of time needed to develop this kind of work rather than the short 30-40-minute periods which give too little time for individual or group work. But it is not impossible for him to work in this way even in a school that is not planned with this approach in mind. Similarly, although, as we shall see, there are considerable advantages to be gained from organizing much of this work on a team-teaching basis, the individual teacher whose school does not provide him with such an opportunity need not feel that there is nothing he can do himself. The practical suggestions which follow, although directed primarily at those teachers whose work involves them with mixed-ability groups, should be seen as having relevance for all teachers regardless of the type of teaching situation they are in.

Content

Our main concern is with the techniques of the individual or group assignment and the first question that arises is that of how to decide on the content of such assignments. As in most areas of educational decision, one can find almost a complete spectrum of answers to this question, from those who would say that content is not affected and that it is method only that is changed, so that common syllabuses for all pupils will still need to be planned in advance, to those who would place no limitation at all on the activity of pupils but would allow them complete freedom of choice in what they wish to do and learn. The answer that each teacher gives to this question will depend on the particular circumstances in which he finds himself and on the view he takes of the wider purposes of education and of his own goals within that context. Let us consider the two extreme positions in a little more detail.

The teacher who is required to work to a detailed syllabus will have little choice in this matter anyway. For him the objectives of his teaching are assigned and, while he can gain much from the methodological advantages that the mixed-ability group and the individual or group assignment approach can offer, there is little for him in the opportunities such a set-up provides for rethinking the purposes of education. Similarly, the teacher who may have much more freedom of action than this but whose view of education is such as to lead him to see educational goals in terms of subject-matter will want to preplan a syllabus for his classes, whether in a particular subject area or in a combination of subject areas. At this end of the spectrum, then, traditional syllabuses, perhaps revamped in an attempt to achieve integration of subject areas, will provide an answer to the question of the content of individual or group assignments. It was argued in Chapter 2 that such an approach loses many of the advantages of the mixed-ability set-up, but it is still perhaps the most usual situation, even in secondary schools that have changed to a mixed-ability pattern of organization, and what will be said soon about method is as relevant to this as to any looser situation.

At the other end of the spectrum is the teacher who is left to himself in the classroom and elects to give his pupils a similar degree of freedom. It may be argued that such a teacher is avoiding the issue of content every bit as much as the one who relies on a syllabus. It would certainly be argued by many, again as we suggested in Chapter 2, that such an approach is based on a misunderstanding of education, a conceptual confusion of education with other related but very different notions such as 'growth' and 'maturation'.

However, to argue thus might be to miss the point, since a teacher who approached his work in this open-ended way might wish to reply that, far from avoiding the issue of content, he is interpreting it in a highly individualistic manner, and that, far from confusing education with growth or maturation, he has a very clear view of education as intimately connected with the individuality of each pupil, that in fact he makes not one but many decisions on content as he approves, modifies or disapproves the choices of areas for study made by each of his pupils.

Clearly, much depends on the degree of freedom given to the pupils and the way in which the individual teacher handles the situation; it is not the absence of a preplanned syllabus in itself which lays one open to these charges. One criticism, however, must be applied to this approach generally and that is that it creates a job for Superman. No mortal teacher is, or can be, equipped to deal adequately or successfully with what is likely to come up if thirty pupils of whatever age are encouraged to work on their own in this unlimited way. One has only to think of the endless stream of questions that flows from any small child to realize what is likely to result from encouraging pupils to explore with complete freedom those things that interest them. If we also remember the greater age, experience and range of interests of secondary pupils, the point will be more than clear.

Aware of this problem and also of the opportunities that would be lost if clear statements of content were drawn up in advance, many teachers have chosen the theme approach as one that gives most of the advantages of the free situation but protects them from a crippling variety of demands. A broad area of work is selected and pupils are offered a range of choices within that area. It is felt that this kind of approach has advantages for the pupil also in so far as, if the theme is well chosen, he can see his work in a context which is broader than the immediate task confronting him but not so broad as to be entirely beyond his own horizon. The main purpose of a theme, then, is to provide a framework in which both teacher and pupil can work securely, profitably and successfully.

Choice of the right theme is, therefore, of crucial importance. A theme which is too narrow, 'The Earth Worm', for example, will defeat the purposes of this approach since it will not be a theme at all; it will be a statement of content like a syllabus and will provide little freedom of manoeuvre for teacher or pupil. On the other hand, it is equally unsatisfactory to go to the other extreme and choose a theme which is so general as to provide no real framework at all. A theme like 'Man and His World', for example, is again no theme at all since it is hardly definitive and does not do for teachers or pupils the very thing a theme should do, namely to delimit the area of their work. Some of the dangers of this loose approach are well described by Tom Gannon in his account of the work of Milefield Middle School (1975, p. 78).

> Attempts to counteract the deficiencies or minimise their effect by means of superficial overall themes or topics which supposedly 'integrate' the curriculum can end in shallow performance levels and confusion of aims and purposes by both teachers and taught. This is in no way to denigrate the many excellent themes or projects undertaken by Junior schools, from which we learn so much. In every case, these are most successful when the teacher partners cut the coat according to the cloth, in other words limit the experience to the bounds of the possible. A salutary reminder of the confusion created by over-enthusiasm lies in the answer of the eleven year old who was asked 'Why do you think we are doing these subjects together?', and replied 'Because we are not clever enough to do them separately.'

A good theme will be one which will provide teacher and pupil with both a structure and as much freedom as each can tolerate.

To provide a coherent structure for the work of teacher and pupils, a theme must be a theme in the full sense of the term. There must be a unity to all that is studied under the label of any particular theme and that unity must be based on a logical rather than a contingent association of those subjects or areas of enquiry that are being subsumed under this particular heading. If we do not hold on to this as a basic principle, then the idea of a theme loses all point, since there will no longer be any coherence either within the separate areas of the individual pupil's own work or between his work and that of his fellows. To use a theme as the basis of a programme of any kind entails a concern for giving the work of all pupils this kind of coherence and point. To take as a theme 'Hands', for example, and to allow this to be explored through biology (the physical structure of hands), industrial sociology (working with one's hands) and religious studies (the 'laying on of hands'), or to allow the theme 'Water' to lead to a study of H_2O, water transport and watermelons is to miss the point of this kind of approach to education and to introduce a random element and a lack of coherence — or worse, a false and misleading semblance of coherence — into pupils' work that is hardly consonant with what most of us understand by education. The main purpose of a theme, then, is to provide an intelligible, coherent and logical context for the work of all pupils.

Several further factors must be borne in mind when a theme is being chosen. The ages, interests, aptitudes and abilities of the pupils concerned must clearly be a prime concern. A theme must provide pupils with scope for work of a kind they can cope with and profit from. A second important consideration must be the competences of the teacher — or of the team of teachers, if this approach is being used in conjunction with team-teaching. It would be very foolish of a teacher to select a theme which required him to break new ground on almost every front to cope with all that pupils might want to work on within the theme and to take advantage of the opportunities it might offer to further their education. Thirdly, a theme should be chosen with a clear view of the goals that it is hoped will be achieved by it or the principles that it is intended shall inform the work of every pupil. A theme is essentially a vehicle for teaching and should not be regarded merely as a means of keeping idle hands busy, although this is, of course, another important consideration. Finally, regard must be paid to the local conditions that prevail in the school and its neighbourhood. There is little point in selecting a theme which will generate a lot of practical activity in a school which is very short of workshop facilities nor in undertaking to work on 'The Lower Thames and its Environs' with a group of pupils who live in Glasgow. All of these factors and many others that will quickly spring to the mind of the experienced teacher must be kept much to the fore when selecting an area of enquiry for a class of pupils.

Once a theme is chosen, the next decision that must be made is whether all the pupils will work on the same things and engage in the same activities or whether they will be allowed to choose their own field of study and devote their attention to that. It may be felt, for example, that out goals will best be achieved if every pupil is required to look at the theme from a number of different points of view and to experience all or several of the activities planned. Some of the Nuffield science projects are good examples of this kind of approach. A theme such as 'Mass' might have as its aim the development of an understanding of several related concepts such as those of weight, volume and specific gravity. If this is the intention, then it might be argued that it will best be achieved by taking each pupil through a 'circus' of experiments, which enable him to experience the problems of weighing and measuring a variety of solids, liquids, and possibly gases also and to examine the relationships between the data he thus acquires. Similarly, a history teacher might set up an exploration of Norman England by requiring all pupils to work in turn on the details of the Battle of Hastings, the armour and weapons in use at the time, the castles, the Domesday Book, monasteries, feudalism, the open-field system and so on, and by arranging for these separate enquiries to be related to each other in such a way as to build up for each pupil a clear picture of the age. In this kind of situation, where the 'circus' approach is regarded as the best method of achieving one's aims, this stage of preparation requires the careful planning of each of the activities or areas of study seen as essential to the achievement of the whole and of the means of bringing them together. Once this has been done, the problem of content has been solved.

However, it may be felt that what is needed is to encourage each pupil to pursue in depth only one or two aspects of the theme, to share his findings or the findings of his group with the others in his class and to see other aspects of the theme by looking at the work of other individuals or groups. A theme such as 'Communications' — a very popular theme in situations were an interdisciplinary approach is being deliberately fostered — is one where it is often felt that children can be helped to see the point and can come to understand the importance and relevance of communications in human development by each exploring one or two aspects only, provided that they are given plenty of real opportunities to share their findings with each other and can thus see the many different sides to the concept.

Some may be led by their interests into approaching the topic from an historical point of view and working on the history of the development of roads, canals, railways or sea-transport; others may prefer to consider the geographical problems that are involved; others still may be drawn to an exploration of the scientific aspects of communication, the development of the telegraph, telephone, radio or television, as well as of various forms of mechanical propulsion; another group may wish to consider some of the implications of improved standards of communication for man's moral and social life; and there is much scope for yet another group to consider language as a form of communication or non-verbal communication through the arts.

In this way, each pupil can have the benefit of working to some depth in the area that is of most direct interest to him while gaining some view of the breadth of the theme by seeing what his or her colleagues are doing.

Clearly, this approach creates a much looser situation than the 'circus' approach and the teacher has now many individual decisions to take concerning the proposed content of each pupil's or each group's line of enquiry. Many lines he will be able to foresee and to think about and plan for in advance, but it is always possible and often happens that pupils will themselves come up with suggestions that are as good as any the teacher has thought of and these will need to be assessed on the spot.

On what criteria should decisions of this kind be made? In the first place, we should not forget that one of the main arguments in favour of this kind of approach is that working through pupils' own interests brings great gains in motivation and that the best way of getting pupils to work is to allow them to work on things that interest them. Psychologists, such as Piaget, who have concentrated their attention on a study of intellectual development, have stressed the importance of intrinsic motivation, of work done for its own sake rather than for some extrinsic reason, in encouraging and promoting intellectual development. This is clearly also the kind of motivation we should be seeking if we are concerned with education in the full sense of the term, since to be educated one must have come to value the content of that education for its own sake. Intrinsic motivation, however, must be sought in the pupil and not in the content of what is offered to him or the methods we adopt in our teaching of him. Intrinsic motivation can only be achieved if we allow a pupil to select his own work and to become absorbed in it.

However, much play is made in theoretical discussions of education of the fact that what a pupil is interested in is not necessarily the same thing as what is in his interests (Dearden 1968), that some interests children and young people have may need to be positively discouraged and that education involves more than pursuing hobbies and requires a breadth of experience greater than is encompassed by the interests of the average pupil. The dangers of merely catering for the interests of each pupil, therefore, are firstly that his interests may be such that we would regard the pursuit of them to be miseducative, if not positively harmful, and secondly that, even where this does not apply, merely to pursue interests one already has may be to miss out on many things that we would see as being constitutive of an education in the full sense of the term. This could be particularly to the disadvantage of the pupil whose home provides him with a limited cultural background and, therefore, a restricted range of interests.

It must be stressed again, as it was in Chapter 2, that while accepting the motivational value of working through pupils' interests, teachers must be prepared to make firm decisions about what these interests lead their pupils to undertake. The principles that underlie the planning of work even in this kind of free situation must be kept in mind when evaluating proposed lines of enquiry, as must the need for these enquiries to satisfy the demands for coherence we have already discussed. We must, of course, be prepared to consider these principles in relation to what we know about individual pupils. We may feel, for example, that a great deal has been achieved if we have produced even a spark of interest from certain pupils and that they should be allowed to do what they want to do merely on the grounds that they will thus derive the satisfaction of doing something. On the whole, however, decisions of this kind must be made in the light of what we consider to be worthwhile activities and what we consider likely to lead towards the goals or reflect the principles we have in mind.

Furthermore, pupils will tend to be interested in the superficially more attractive aspects of a topic and will need to be led into those aspects which are less exciting but which may be necessary to give coherence to the whole and to promote understanding. If education is to mean anything it must involve the extension and development of pupils' interests rather than the mere satisfaction of them (Wilson 1971). It may be true that what is worthwhile can only be defined in terms of what is worthwhile to the pupil, but if the teacher is to play any positive role in education, it must be by his or her skill at developing the potential of the individual pupil revealed through his interests rather than merely feeding those interests in the way that anyone with an adequate supply of paper, paint and other materials might do. What is being argued is that the pupil has a right to contribute to the discussion of his own education, but that he is not competent to decide entirely for himself its goals, content or methods. It is the failure to realize this and the resulting involvement of many pupils in totally undirected activity that has led to most of the criticisms that have been made of 'free' methods in schools.

Method

With a theme chosen in the light of these considerations, and decisions of content made or at least principles established upon which they will be made, the teacher's next problem is to put all of this into action. We must now turn, therefore, to a discussion of some methods that might prove helpful to the teacher adopting an individual or group assignment approach.

The first task in any educational undertaking is to get the right psychological 'set', to arouse interest, to ensure motivation, so that the teacher must begin by trying to communicate to all of his pupils the interest that it is hoped he himself already has in the project he has planned and to show them that there is something of interest and value in it for everyone. There are several ways in which this can be attempted.

Many teachers favour the 'impact session', the 'key' or 'lead' lesson, as the most effective method of arousing interest. The new work is introduced to the pupils by a presentation of as stimulating and exciting a kind as can be devised. Clearly, there are advantages here in having a team of teachers involved, but there is a great deal that the individual teacher can do with his own class. Films, audio- or video-tapes, film-strips and other such aids will obviously offer enormous advantages. Visiting speakers can also contribute a lot to this kind of exercise — a new face is a stimulus in itself — although care must be taken in choosing visitors, since some can have the opposite effect to that intended. It is particularly useful to consider what parents might be able to offer in this kind of situation, since among them will be experts of all kinds whose knowledge can be tapped — without fee — and this is one way of involving them in the work of their children in a more direct and mutually profitable manner than that offered by a Parent Teacher Association. A programme can be devised for one lesson or several, using devices of this kind to show pupils as many aspects as possible of the theme or area of work the teacher is about to take them into in a way that it is hoped will arouse their interest and make them want to explore further.

A second method of arousing the interest of pupils, which may be applicable to some kinds of theme, is a visit or series of visits to places in the locality that have some bearing on the area of work to be explored. A visit to a local factory engaged in some industrial process that is based on the scientific principles it is hoped the pupils will come to understand, a visit to the docks to see the kind of cargo that is being handled and its origins, a 'nature walk', a visit to the local church or any kind of outing that can be devised to show pupils aspects of the project can be expected to arouse more interest than the most inspired lesson in the classroom. Nor do such visits need to be to rather obvious and special places like the local zoo or the Tower of London. A moment's thought will reveal to the imaginative teacher what he can bring to the notice of his pupils even on a walk around the vicinity of the school, if such a walk is carefully planned. Much can be done to arouse interest in certain kinds of theme by this kind of planned outing.

A third way of stimulating the interest of pupils in the work about to be undertaken is to surround them with examples of what they could do and perhaps also with the material and equipment they might be working with. This kind of technique has proved highly successful in many infant and junior classrooms where teachers have long known the value of creating a number of displays, each related to some particular interest or activity — measuring, weighing, number, reading and so on — and allowing children to be stimulated simply by looking at these displays in passing, as it were. An exhibition is often seen by teachers as a kind of end-product of work on a theme of the kind we are describing, but many of them, when they have got there, have discovered that for the pupils, rather than being a goal, the exhibition is a new starting-point. A display or exhibition can be a good way of beginning a project and a source of great stimulation for the pupils. Similarly, to be shown the materials available can also have this effect. Again, infant and junior teachers have long recognized this and known the advantages of making available to children sand trays, water containers, Cuisenaire rods and the like and allowing them to handle such apparatus informally, to 'play' with it. One can imagine secondary pupils being equally excited if, on entering a science laboratory, they found the apparatus, equipment and materials laid out and were allowed to browse among it for a while. To surround pupils by visual and other examples of what one has in store for them may be, therefore, as good a way as any of arousing their enthusiasm.

One final point needs to be made on this. Teachers should not be above manipulating such approaches to suit their own ends and purposes. Whichever method is chosen, it is worth remembering that there are advantages in stressing the less obvious aspects of the theme, since the pupils themselves will immediately see the more obvious ones, and in stressing those that for one reason or another the teacher wishes to promote. It is at this point, at the very outset, that we begin to make use of the interests of our pupils to forward their education.

If we have been successful in presenting the project to our pupils in a stimulating way and arousing their interest, the result of our efforts will be a class of pupils, eager, or at least willing, to get to work. Our next problem is the organizational problem of getting them started on their individual or group assignments. As far as possible, this should have been planned well in advance. No matter how great the interest we have aroused, it cannot be maintained for long unless we are ready to follow it up by providing pupils with the wherewithal to begin their work. If we are engaged in the kind of project where the syllabus has been pre-planned, then it is a matter merely of good classroom organization and of having clear instructions, perhaps in the form of work-cards, and the necessary materials ready to be able to get them all down to work very quickly. The same is true if we are adopting a 'circus' approach. We may be prepared to allow them some freedom in choosing where they will begin or which group they will work with, but there is nothing really complex about getting this kind of thing off the ground.

Once again it is in the freer situation that the problems are most acute. If we are really prepared to consider suggestions from the pupils, this can be a very lengthy business and without careful organization can result in chaos. It must never be assumed that because the pupils are to be allowed to make their own choices the teacher must not preplan. Without very careful and extensive preplanning disaster will inevitably ensue. It is not difficult for teachers to foresee most of the lines of enquiry that children are likely to come up with and, as we have said, the initial programme should have been framed to promote interest in the kind of thing the teacher is prepared for. And so many pupils can start work right away, even in this free kind of atmosphere. Again work-cards may be useful in providing initial guidance and the teacher must see that the resources and materials needed are ready to hand. If most can be started off in this way, the teacher is free to deal with the other cases, those with suggestions that were not foreseen and which will need careful thought and evaluation and the inevitable group of pupils who are not stimulated, inspired or even mildly interested and will need to be privately and individually stimulated or, in the last extreme, directed into something, if they are not to constitute a disruptive influence on the work of others.

From this point on, the teacher's job is to keep a careful watch on the progress of each pupil and each group of pupils. He must make sure that his pupils are really working and getting full educational value from what they are doing, that they are working with understanding and not 'going through the motions' without thinking about what they are doing. He must ensure that as far as is possible they have the materials and resources they need. He must see to it that the initial impetus of interest is maintained as long as is possible and added impetus provided as and when it seems necessary for individuals or for the whole class. He must be on the lookout for opportunities to extend the range of their work by leading them on from the point where they started to further developments that, left to themselves, they would not have contemplated or to turn them in the direction of something new if they appear to have exhausted the vein on which they started. And all the time he must be doing this against the background of the goals or principles he began with, modified perhaps in the light of his continuing experience of what has turned out in practice and always with consideration for the individuality of each pupil. All of this is, or should be, part of his professional skill as a teacher and can be left to that.

However, there are certain general points that he needs to keep in mind as he guides his pupils' work in this way. To begin with, he must not lose sight of the danger, already referred to, that this kind of approach can lead to undirected and haphazard learning and can degenerate into activities that have no real educational value. At the same time, his own role can be reduced to that of a storekeeper providing materials for pupils to use but no guidance as to how they should use them to further their education. It should not be the intention to set up a kind of hobbies club in which pupils do what they want and teachers provide them with the materials they need and help out with the difficult bits. This is not the object of the exercise at all. The teacher should be doing as much, if not more, teaching in this kind of situation as he does in a more formal lesson and the pupils should be working as hard, if not harder, since they should be engaged on tasks which stretch them and which make more demands than anything which they would choose to do at home as a pastime. This approach is, after all, a method of teaching, not an alternative to teaching. I well remember seeing an exercise of this kind in which one particular boy was engaged in some work on cars. He was making a folder or scrapbook containing pictures of different makes of car and listing the specifications of each. The teacher — in this case a student — was standing at the front of the class handing out paper, scissors, glue and so on in whatever quantities were needed — so enthusiastically, in fact, that there was later a stationery crisis in the school. It was clear that the boy had no understanding of the figures he was listing (for example what 'cc' means and what it signifies in relation to a car's performance), and that the teacher was making no attempt to take that boy nor any other from the interest he had shown towards something more demanding. Except for the free paper and other materials, the boy was doing no more than he could have done at home.

A second danger of this kind of approach is that it can easily lead to a concentration on fact-finding exercises to the detriment of other activities, perhaps of a more creative kind, that are equally important elements in education. It is easier to view enquiry, 'finding out', as something that results in the acquisition of a lot of propositional knowledge than to see that there can be other methods of exploring the world about us — through art, craft, dance or drama for example — that may not result in propositional knowledge but can certainly bring understanding of a different kind which is just as important. Indeed, this is a criticism not only of this new kind of approach to teaching, but perhaps even more of the traditional English secondary curriculum which has placed great emphasis on the 'factual' subjects but often seen little value in engaging pupils in any kind of creative activity. This has resulted in a very one-sided kind of upbringing for many children and for this the theorists must bear as much responsibility as the curriculum planners themselves.

Interest-based, enquiry-based teaching, however, offers real opportunities for creative work of all kinds — creative writing, art, model-making, dance, drama and so on — but it is easy to lose sight of this fact and to allow scope only for the 'factual' subjects. There is no theme that does not have this kind of dimension to it if teachers are aware of the need to look for it. A theme on 'Women in Society', for example, undertaken with classes of secondary girls by a group of students on a teaching practice, led not only to a sociological and historical examination of the topic but also to a visit to the National Portrait Gallery to see how women have been portrayed by artists which had direct results in the creative work of some of the pupils. Even a theme as apparently 'down to earth' as 'South Yorkshire and its People', undertaken by a member of staff at Northcliffe Community High School, was approached in such a way as to encourage the production of a good deal of creative writing alongside the geographical and sociological (Barrs, Hedge and Lightfoot 1971).

The teacher who is aware of the danger of allowing such projects to develop into fact-finding exercises and of the need to promote other kinds of activity must work specifically for this, and lead pupils into it by means of what he presents to them, what he encourages them to follow up and the emphases he gives to the work. In using this approach with teachers in training, we quickly became aware that the educational diet of a potential teacher has normally been such that when asked to do any piece of work he reaches for a file and starts to collect 'facts'. On one occasion, therefore, we expressly forbade such activities and insisted on a nonverbal presentation of their findings. The results were most interesting and there is no doubt that we achieved our aim, which was to show them that there is as much educational value to be gained from that kind of activity as from most of the things that traditionally go on in schools and colleges.

Two examples

Many of the points I have tried to make about the approach to teaching and learning through individual and group assignments are illustrated very clearly and cogently by two projects undertaken at Milefield Middle School and described by Brian Walker, the school's Art and Craft Adviser (1975, pp. 101-109).

> I now wish to go into detail concerning two major year group projects. One involved work and study of one of the largest and most architecturally beautiful houses of the eighteenth century period, namely Wentworth Woodhouse, near Rotherham, the Great East Front of which is at present a College of Education. The project was organised in collaboration with the college staff and a small group of students. The other was a project entitled 'Man and His Materials'. The aim here was to visit small firms in the Barnsley area, observe such materials as clay, wood, glass, coal and metal being used in different processes to produce commercial goods and then to use the same or similar materials in school either scientifically or aesthetically. The resultant school work would emphasise the various qualities of the materials and their creative possibilities. I use the word 'creative' to envelop scientific discovery as well as aesthetic expression.

> Both projects were carried out with the same children. The year group was relatively small, ninety children, so new types of organisation were possible and all the areas in and around the year group space were utilised to the fullest extent.

The Wentworth project

General Aim: To study the Wentworth Area, the House, the Village and their geographical, historical and sociological relationships to one another. The work to be appropriate to twelve to thirteen year old children.

Specific Aim: To explore and discover form, construction, style, age, location and relationship of buildings and areas within Wentworth as a means of achieving certain educational objectives.

Objectives:

1 To increase the children's awareness and enjoyment of their surroundings, people, things indoor and outdoor, by accurate observation of this new environment.

2 To increase their understanding of man-built elements of the landscape.

3 To stimulate growth of children's:-
a) Critical faculties.
b) Ability to apply their discoveries and analysis to future projects and subsequently life outside school.

4 To provide opportunity for imaginative and self-expressive work, oral, written and visual.

My task as Head of Year Group was to make several preliminary visits to Wentworth to discuss details of the project with college staff beforehand. I also needed to weigh up all the possibilities. Arrangements were made with local notables including the Headmaster of the village school, who in fact agreed to show us old record books and even cordially invited us to use the school in case of bad weather, the local vicar, who agreed to open the old church (now in a dilapidated state) and put on a bell ringing ceremony in the new church, and the Estate Manager, whose co-operation we needed to allow us to look over the estate, to talk about forestry and to act as guide. Other local people were also consulted. I then co-ordinated all aspects of the project.

As this was my first effort at organising such a project, I was naturally anxious for it to succeed and was therefore dissatisfied until every minor detail had dropped into place. Year group meetings took place so that my colleagues could suggest ideas and methods of organisation. It was important too that advisory staff be encouraged to contribute.

The children were split into mixed ability groups with at least one boy or girl who had potential as a leader. It was also important to give the less able children a certain degree of responsibility. Very careful consideration was given to which children each group contained. Within a year group situation, it is vital that every child is known thoroughly — perhaps more so than in any other kind of grouping.

Several visits were organised in advance and for the purpose of these the children were arranged conveniently into six groups of fifteen, each with a member of staff or students in charge. The groups were as follows:-

Group 1 To study the House interior and the history of the family.

Group 2 To study the exterior of the House, its architecture and immediate grounds.

Group 3 To study the two churches in the village — one eighteenth century and the other twelfth century.

Group 4 To study the village — mainly one long street.

Group 5 To study the wider surroundings of the House, the Estate and the pastoral industries carried out.

Group 6 To study the geography of Wentworth, the various monuments and follies erected by the late family, to seek a relationship between the House and the village, and gain as much information as possible from local people.

Staff and students were allocated to groups according to their individual skills and interests.

One of the college lecturers, an authority on the history of Wentworth and the House, began the project by giving a key lecture to the whole year group. The various rooms along the Great East Front were mentioned, the names of notable members of the Fitzwilliam family were made known, several anecdotes were passed on and slides were shown.

Each child was provided with a map showing the ten mile journey from school to Wentworth, a questionnaire to answer en route, five detailed sheets on the Fitzwilliam family, the House and the rooms inside. Everyone was also provided with writing and drawing materials, and a board and clip. Some children brought cameras.

1st Visit: Groups 1, 2 and 5 tour of the House and grounds at half hour intervals.
Groups 3 and 4 tour of the village, church bell ringing.
Group 6 visit to small farm owned by local builder.
Groups 1, 2 and 5 alternating between conducted tour of house, walk along terrace and sketching East Front.
Length of visit — 2¼ hours.

2nd Visit: Groups 3, 4, 6 to visit House.
Groups 1, 2, 5 visit to Estate woodyard, through village.
Collect samples.

3rd Visit: All groups to begin work on assignments.

Subsequent visits by individual groups in school minibus to reassess, answer new questions, refresh observation.

The main bulk of the work was carried out in school on two afternoons per week, although nothing was rigidly fixed and staff used other periods of class time to carry out written work, painting, model making or display. Latterly, the afternoon timetable was abandoned altogether in a last ditch effort to complete as much of our anticipated work as possible.

The whole project was brought to a climax by a selected display in the school's central Display Area. Selection had to take place. There wasn't enough space to house all the work.

The real success of the adventure may never really be known. There was no doubt that the children, coming from a mining community into a rural setting were deeply impressed by the splendour of such a magnificent House, its chandeliers, mirrors, libraries, wall decorations, pillars and so on. Simply to be there was experience enough.

It would be pertinent to observe a few examples:

1 David, aged fifteen, shy, introverted, quiet to the point of being silent, was greatly impressed by the stone balustrade around the edge of the roof along the Great East Front. He moved along the 600ft. length of the building sketching the different skyline shapes. Back at school, his attempts to recreate his sketches in pen and ink failed. I helped him by suggesting that his drawings might be better as silhouettes either inked in or even cut out of black paper. He was pleased with his first new attempt and continued to produce dozens of these. Any attempt on my part to suggest a change of approach or something different met with blank rejection. He had found something that was pleasing him. On the wall, the silhouettes showed up in a remarkably striking manner almost echoing the building itself.

2 Stephen and Jeffrey usually spent most of their class time gossiping and being generally lazy. It was quite a revelation to find them unusually engrossed in the making of a model of the huge portico, Corinthian columns and all, from card, wood and polyfilla. Although I kept a distant eye on them working out in the year space, they needed little help and completed a lifelike model which we photographed. Stephen has left school now. His model has pride of place at home. His parents helped to transport it home safely.

3 Neil, aged thirteen, an asthma sufferer who found enormous difficulties in living a normal life, was one of those relatively few children who showed a rare talent for drawing, especially portraits. His picture, a large shaded pencil drawing, illustrated the dream he had had about Wentworth. The drawing contained several elements closely interrelated and overlapping. These included the House, the marble statues inside, the village church spire showing above the trees, a domed folly and the Fitzwilliam coat-of-arms. It had immediate appeal to the other children. An on the spot lesson ensued, as I used Neil's picture as a lead into the History of Art and the Surrealist Painters. Wentworth, naturally presented many such opportunities to discuss Art and Architecture. Fortunately, the school library is well stocked with books on both subjects.

Man and his materials

For this project, the general weekly timetable was abandoned for the whole of the first week so that all the visits could be made. This was essential so that work could begin immediately visits to firms were over.

Visits to the following industrial works were arranged.

1 The Coalite Coke and Chemical Plant nearby (coal, chemicals). A high proportion of the children's fathers work here.
2 The local colliery brickyard (clay).
3 Glass Works, five miles away (glass).
4 Steel Works, six miles away (metal and machinery).
5 Shoddy Mill, six miles away (fabrics).
6 Pipe Works, eight miles away (stoneware clay).
7 Motor Dismantlers, local (scrap metal).
8 Wood Yard, ten miles away (various timbers).
9 Local woods and plantations (timber, natural science).
10 West Riding Environmental Studies Centre (farming, animals).
11 Film and lecture at the school by local newspaper PRO (paper and printing).

. On this occasion, the children were given complete freedom of choice as to which visits they wished to make. The criteria for the choice were (a) the materials they most wanted to work in and (b) two visits each, plus the Coalite Plant and the lecture on production of newspapers.

My first expectation was that all the children would eagerly seize the opportunity to visit the glass and steel works, seemingly the most exciting of all the visits arranged.

The majority of children, however, asked to visit the school farm and the woods. Hasty negotiations took place between myself and the children in attempts to balance the groups. Second and third choices were made which solved the problem.

Again, the task of contacting the various firms and arranging the visits was my main occupation. I must say that the amount of co-operation was delightfully encouraging. If I may jump a paragraph or two, one firm actually laid on tea and biscuits for the children and provided them with leaflets and brochures.

Being an artist, I arranged six groups and named them Red, Orange, Yellow, Green, Blue and Purple. Mathematics and Art are so easily reconcilable.

Group	1st Visit	2nd Visit	
Red	Brickyard Thursday 14th 2.00 p.m. On foot	Pipeworks Wednesday 20th 2.00 p.m. By Bus	Clay
Orange	Glassworks Friday 15th 2.00 p.m. School Mini Bus	Pipeworks Wednesday 20th 2.00 p.m. By Bus	Glass Clay
Yellow	Woods Tuesday 12th 1.30 p.m. School Mini Bus	Environmental Studies Centre Thursday 28th School Mini Bus	Timber Soil etc.
Green	Steelworks Wednesday 13th 1.30 p.m. By Bus	Brickyard Thursday 14th 2.00 p.m. On Foot	Metal Clay
Blue	Steelworks Wednesday 13th 1.30 p.m. By Bus	Scrap Yard Thursday 14th 11.00 a.m. School Mini Bus	Metal
Purple	Shoddy Mill Wednesday 13th 1.30 p.m. School Mini Bus	Wood Yard Tuesday 19th 2.00 p.m. School Mini Bus	Fabrics Fibres

All groups to Coalite Plant Monday 11th 10.00 a.m.-3.45 p.m.
All groups film/lecture "How a Newspaper is Made" Tuesday 10th 10.00 a.m.

Aims:

1 To use the local environment in a fresh way following on from the aims and objectives of the Wentworth Project.

2 To discover fresh outlets and possibilities for further school projects.

3 To provide a six weeks' intensive working atmosphere.

4 To encourage staff to work closely together as a team providing a blanket of information and guidance.

5 To force the children deliberately into a situation whereby they would need to seek help, advice and information from whichever members of staff could provide it. No one member of staff was in charge of any group. Staff simply went on the visits they most fancied.

6 To work in a practical rather than theoretical way. Each child was, however, expected to write an account of his visits and produce an informative booklet about his chosen material.

7 To use the chosen materials intelligently through a discovered knowledge of their various properties.

8 To use the year group areas and specialist rooms to their maximum extent.

M.A.G.—F

Spaces used:

1 The fourth year group classrooms
2 The year group space
3 The Science Area
4 The Workshop
5 The Home Economics Area
6 The Library and Audio Visual Aids

The total cost of the journeys was no more than 15p per child. The variety of work produced was most encouraging, highly creative and totally absorbing.

Peter, aged thirteen, had had little interest in school work. He could not read, was ridiculed by his peers, had enormous difficulties at home. His mother, the driving force in the family, had died the previous year. His trust in one member of staff, individual attention, a six weeks' course in outdoor pursuits organised within the year group, and more, helped Peter with his reading and enabled him to become more socially acceptable. He turned out to have an enormous sense of humour. As part of the "Man and His Materials Project", he not only carried out various experiments on coal — heating the coal in a bunsen burner and extracting tars etc., — but also constructed a blown glass mobile, each shape containing brightly coloured liquids. The mobile was placed in the year space in order to catch the sunlight from the window.

Work produced included:

Glass blown sculpture
Metal sculpture — copper, steel and aluminium
Pottery — earthenware and stoneware
Beaten copper work
Wood turning on the workshop lathe
Functional objects in wood and needlecrafts
Fabric collage
Ceramic sculpture
Etching and various forms of printmaking including silkscreen
Germination and propagation of seeds and plants
Wood carving
Experiments with glaze and other ceramic materials
Batique work

Problems of particular subjects

Finally, a word must be said about the difficulties teachers of certain subjects might have in trying to adopt this sort of approach. So far, it has been assumed that the subject involved is not a significant factor, but there are many who feel that the nature of some subjects is such as to preclude the possibility of approaching them in the ways outlined. For example, it is said that certain subjects require a 'linear' approach, that certain kinds of knowledge must be developed step by step and cannot be tackled by starting from whatever point happens to offer a link with some interest the individual might happen to have. In particular, this kind of assertion is made about the teaching of foreign languages, mathematics and science and many teachers who specialize in these subjects will claim that any attempts to approach the teaching of their subjects via individual assignments would be sure to lead to disaster. On the other hand, there are teachers of these subjects who have welcomed a change to mixed-ability groups, have happily given up relying solely on class-teaching techniques and would claim no losses and some positive gains from their achievements as a result (Clayton 1971; Hytch and Tidmarsh 1971; Bosworth 1971; Hamilton 1971; Warnes 1971; Prettyman 1975; Haslam 1975; Walmsley 1975).

What is the truth of the case? Once again, it would seem that we must look more closely into each situation. To some extent, the disagreement arises more from different conceptions of the subject or from a failure to distinguish between the different purposes that one might have in teaching different aspects of the subject than from different evaluations of the effectiveness of the method. Different conceptions or different purposes will lead to different emphases within the teaching of a subject and it will be clear after a moment's reflection that while it may be true that some aspects of the teaching of these subjects, such as the teaching of basic French grammar, may require a linear approach, other aspects, such as the development of an awareness of the history, culture, economy and so on of the French nation, can be dealt with, and perhaps should be dealt with, by the kind of interest-based, enquiry-based methods we have been discussing. Furthermore, even the teaching of the basic 'facts' or skills may in some cases be better done by means of the individual assignment, as in the case of some of the Nuffield science projects already referred to, although here of course the aim will be not to encourage enquiry into an area of the pupil's own choice but to require each pupil to tackle a 'circus' of activities within a prescribed area or to acquire some basic skill.

Many teachers have also discovered that a linear approach even with streamed classes requires a great deal of individual help and that in the end the individual assignment is necessary here too. This was certainly my own experience when I was teaching mathematics some years ago to a third-year 'B' stream class of more than forty boys in a secondary modern school. The range of abilities was so great — some boys able to cope with elementary algebra, others having difficulties with the most basic calculations — that individual assignments were the only way of ensuring some value for all pupils and of retaining my own sanity.

It may be felt necessary, however, in all subjects, particularly those where it is claimed that a linear approach is crucial or those where the teacher is more at home working in a more formal way, to arrange for some straight teaching of a subject to be undertaken. Certain basic skills may well need to be provided for in this way. If this is to happen, the timetable must allow for it at set times in the week. We shall see in discussing the teaching of pupils with learning difficulties in the mixed-ability class that there is a need to provide special 'remedial' periods for them and for all pupils. A similar need may exist to provide periods for the linear teaching of certain subjects or for the work of certain teachers, and this can be met in the same way. All groupings within the school, including the class groupings, must be flexible enough to allow for the different kinds of provision that will need to be made to cater for individual pupils and individual teachers (James 1968). This is a point we shall return to in Chapter 10.

Summary and conclusions

In this chapter we have looked at what appear to be the two most important aspects of an approach to teaching through the use of individual and group assignments. First an attempt was made to delineate some of the criteria that teachers need to refer to in deciding on the content of such assignments and then we went on to consider some of the methods that they might adopt in organizing their teaching along these lines. Finally, we looked briefly at some of the problems that it is said face teachers of certain kinds of curriculum subject when this approach is adopted and suggested some solutions to them.

In all that has been said so far we have had in mind the individual teacher working out his own personal salvation in his own class or classes and even within his own subject area. In short, we have been concerned to offer advice that any teacher could see as applicable to his own situation. It will be clear, however, from what has been said that many of these developments are likely to be easier to implement or more effective in practice if certain changes are made in the organizational structure of the school. In particular, blocks of time rather longer than the single period are needed to allow for the kind of absorption in a piece of work that this approach is designed to encourage, and collaboration between teachers of different subjects can add another dimension to this kind of teaching and, indeed, protect individual teachers from having to meet the varied demands on their knowledge that it has been suggested some themes may generate.

Many schools have, therefore, altered their timetables to provide blocks of time for this kind of activity and sometimes also to allow for team-teaching. Sometimes this has been done in conjunction with a change to a mixed-ability pattern of organization, sometimes in place of such a change, since, as we shall see, team-teaching can offer mixed-ability situations even in a streamed school. We must now turn, therefore, to a consideration of developments of this rather more extensive kind.

CHAPTER 4

TEAM-TEACHING

In the previous chapter, an attempt was made to indicate some of the ways in which the individual teacher can alter his own approach to teaching towards the more individualistic methods that seem necessary to the teaching of a mixed-ability class. It was shown that a great deal can be done by the individual teacher and that the adoption of a mixed-ability form of organization does not necessarily entail a move to some form of team-teaching. Conversely, it is also the case that team-teaching can be undertaken in a streamed school and the streamed divisions can be maintained within it. In short, mixed-ability groups do not require team-teaching nor does team-teaching require mixed-ability groups.

However, in practice the two are often found together. In particular, team-teaching is usually found in schools where there has been a change to mixed-ability groups and is sometimes used as a means of effecting such a change, since it is possible to introduce some mixed-ability working through a team-teaching programme even within a school whose basic organizational structure is a streamed one. No discussion of mixed-ability teaching would be complete, therefore, without a full examination of the practicalities of team-teaching.

Some advantages

There is little doubt that team-teaching, properly organized, planned and executed, can give an added dimension to the kind of teaching we discussed in Chapter 3 (nor that, if not properly organized, it offers more scope for chaos than most teaching methods). A general aim of this kind of approach to teaching, as we saw in Chapter 1, is variety and flexibility and team-teaching can increase both. There is more scope for working through pupils' interests, since we no longer have to decide either to limit their range to the area of competence of one teacher or to expect that teacher to extend his area of competence to the point where in becoming wide it will also become dangerously thin, where he may become a jack of all trades and master of none. A team of teachers will bring many areas of competence and expertise into play and no member of it need surrender or compromise his specialisms. Indeed, he may in this situation be able for the first time to make full use of skills and knowledge that previously he had little or no scope for, since an interest-based approach, as we have seen, may allow for new developments in content beyond the traditional school subjects. A pupil may wish, for example, to undertake some filming or still photography and this may be seen in its context as a valuable thing for him to do, in which case a teacher who has some skill and experience in this field would be a great asset and would find a new dimension to his work. The kind of individual assignment discussed in Chapter 3, therefore, becomes easier to provide for because of the range of knowledge and expertise that a team of teachers can bring to bear on it.

Team-teaching can also provide a useful flexibility of groupings, since with a number of teachers available pupils can be divided into groups of varying sizes according to the needs of any situation. Sometimes much time and energy can be saved by taking the whole group together for a presentation, lecture, visiting speaker or other such event that is felt to be appropriate and valuable for all; at other times, divisions can be made into groups of normal class size or into smaller groups of, say, four or five pupils for tutorial or other purposes, again according to need.

In Chapter 3 stress was laid on the importance of the right kind of initial presentation of a theme to pupils. Without the motivation that this is designed to generate, little can be achieved. The same is also true of any later occasion when it is felt desirable to have some kind of presentation to all of the pupils together, either by a member of the team or a visitor. This kind of situation is more prone than any other to misfire, to promote boredom rather than enthusiasm, to end in chaos rather than purposive activity. The existence of a team of teachers, some of whom will be better at this kind of thing than others, is again a safeguard against disaster. Those who evince a particular talent for the large public performance can take responsibility for all such presentations, while the other members of the team concentrate on those activities that they are good at.

A related point that should not be overlooked is the scope that a team-teaching scheme offers for providing support for those teachers who, for a variety of reasons, may need it. Every school contains staff members whose contribution to the education of its pupils will be enhanced by the support they can gain from membership of such a team. But this is of particular advantage to the student teacher or the probationer since working with a team of more experienced colleagues will provide exactly the kind of help and advice the young initiate into the profession needs and is entitled to receive. Team-teaching offers the ideal context for a proper programme of induction.

A final point that must be made in listing some of the advantages of team-teaching is the fact that it gives teachers greater opportunities for discussing their work with each other. Indeed, one might go further and stress that it actually requires of teachers that they discuss their work with their fellow team members. Too often there is not enough dialogue between teachers. The individual teacher in his classroom is like a goldfish in a bowl, cut off from others and from the world outside. Some informal discussion often does go on in the common room, but there is nothing like a common task for generating real discussion between people. A team-teaching assignment makes it impossible for teachers to avoid constant and rigorous debate and reappraisal of the principles underlying their teaching, the content of their work and the methods that will best enable them to achieve their goals. Some would even claim that team-teaching leads to better preparation of work, since the work of each individual is open to the scrutiny of his colleagues and inadequate preparation lets the team down as much as the individual himself. Again, team-teaching is not, or should not be, essential for constant and rigorous consideration by a teacher of his professional concerns nor for adequate preparation of his work, but there is no doubt that it provides additional incentives to both.

What is team-teaching?

Team-teaching requires the collaboration of two or more teachers on a common educational purpose (Shaplin and Olds 1964; Lovell 1967). This is the one common denominator of all team-teaching situations. Beyond this there is quite rightly infinite variety. Teams vary considerably in size. Logically, of course, they cannot contain fewer than two teacher members; seldom do they contain more than eight, although there is no reason, other than perhaps certain administrative considerations, why they should not. Clearly, the size of the team will depend on several factors, but the key factor must always be the number of pupils that is being catered for. It would be unrealistic to plan for a teacher-pupil ratio that was very much better than that allowed for in the overall staffing of the school. One teacher for each class of pupils involved in the exercise, therefore, would seem to be the general rule. If an extra member can be made available, then this is all to the good.

Similar variation can be found in the sizes of pupil groups. Some schools have undertaken team-teaching with single classes, although obviously such generous staffing has to be paid for elsewhere in the timetable. At the other extreme, there are occasionally team-teaching programmes that involve the whole school. However, such work is most commonly undertaken with year groups, divided into two if they are particularly large, so that the optimum size for most purposes would seem to be three or four classes, 90-120 pupils.

It is also most usual to find this kind of approach being used with the first and second years in the secondary school, although many have found it valuable in the final year, particularly with pupils not preparing for public examinations, when these are being taught separately from their colleagues, and some schools have seen it as an essential element in the teaching of every year throughout the school. It is also increasingly common to see forms of team-teaching in primary schools. In fact surveys have revealed that it is used in all kinds of school and, indeed, at all levels of education, including colleges and universities, although it is not always specifically called team-teaching (Shaplin and Olds 1964; Freeman 1969).

There is no one answer to the question of how much time should be allocated to this kind of work, since almost every variation can be observed. As so often is the case, it is a matter of adapting to local conditions. Some certainly would be prepared to devote all of the working week to a team-teaching project, but it does seem important to remember that many things that might be regarded as vital may not be adequately dealt with in such a programme — the basic skills of reading, number and learning a foreign language, for example, and remedial work of all kinds — so that it is probably safer to see this as only one aspect of a total curriculum, not as a curriculum in itself. Furthermore, it is always possible to have too much of a good thing and it is difficult for any of us to sustain an interest, however absorbing, from Monday to Friday unrelieved. Most schools, therefore, allocate only a part of the week to work of this kind. Some schools, especially primary schools, allot every afternoon to it; others have given it a day like the primary 'integrated day'; yet others have found three or four half-days per week for it. In practice it is a matter of extracting what time is possible in the face of the many other competing interests and the general consensus seems to be that three half-days per week, ten to twelve periods, is quite adequate and possibly sufficient for most undertakings of this kind.

The last important variable is the range of work that a team-teaching programme is to cover. Team-teaching is used both in situations where a very detailed syllabus has been drawn up and in situations where an enquiry-based, interest-based approach is being used. It is used within particular curriculum subjects and where there are no restrictions as to subject; sometimes in fact it has been used to break down subject barriers. Most often it is associated with a move towards some form of 'integrated studies', although not always with a change to an enquiry-based approach. In practice, the range that a team-teaching project is designed to cover will again depend on local factors, on the proportion of the working week devoted to it, on the subjects that have 'given up' their share of the timetable to allow it to be introduced and on the departments and individuals who are interested in this development and willing to be associated with it. Such local factors most often mean that it becomes a humanities project — there may be a parallel, but separate, team-teaching project in the sciences or they may continue to be taught in a traditional manner. Occasionally, it will be possible for it to cover a full range of subjects. The latter is probably the ideal situation, especially if it is felt desirable to allow the pupils as much freedom of enquiry and exploration as possible, but the important thing is to be clear in advance about the particular situation that prevails, about what should be done and about what can and cannot be done. Any kind of team can work, provided that its members know where they are going, what they can cope with and what they must leave to someone else.

Administrative arrangements

Infinite variety is possible in the type of work undertaken by a team and the type of team that might be developed to deal with it. Whatever kind of team we eventually go for, however, its work can be made or marred by the kind of administrative arrangements made for it. Team-teaching makes greater and more complex demands on the expertise of teachers than class teaching of a traditional kind and we are inviting disaster and being hardly fair to the teachers involved if we expect them to undertake this work without being given, as far as is possible, conditions to work in that, if not ensuring success, will at least not provide them with additional hurdles to surmount.

A first essential requirement and one which applies, as we saw in Chapter 3, to all individual and group assignment work, whether associated with team-teaching or not, is that the timetable should allow substantial blocks of time in which the work of pupils can be developed. This kind of teaching is impossible in single periods of forty minutes or so. If the interest of the pupil is to be maintained, he must be given the opportunity to put in a decent amount of work on whatever he is engaged on each time he comes to it. He must be allowed to become absorbed in what he is doing, to see some positive reward for his efforts and to end each session with the knowledge that he has made observable progress. Some will not be absorbed, of course, or will finish a particular piece of work quickly. These can be moved on to something else. But the pupil who is involved must be given time to achieve something substantial and not be moved from one thing to another, like a rat or a pigeon, by a bell or a buzzer that interrupts him every forty minutes, since in these conditions it is difficult to see how any real habits of study can be developed. It is often said that children cannot concentrate for long periods of time — this has been the justification of the short school period — but we have all seen quite small children 'lost' in something that really interests them for much longer periods than that. It is this ability to become absorbed in something of genuine interest that we must try to capitalize on

Furthermore, the teacher needs substantial blocks of time with his pupils. In a situation where he is not offering them all a short formal lesson but is concerned to work with groups or individuals engaged on different kinds of work, he needs time to get around to all of them, to see where they have reached, to advise on where they should go next, to prod those who are slacking, to make sure that the eager ones do not lose their enthusiasm through a lack of books, materials or encouragement. To ask anyone to work in this way in forty-minute bursts is to invite failure for the project and neurosis for the teacher.

On the other hand, it is equally important not to provide periods of time that are so long that interest becomes difficult to sustain. It is better, of course, to err on this side, since the teacher can always break up the time allotted and vary the programme to avoid boredom. Blocks of time roughly equivalent to three normal periods seem to be the optimum here. This is why many schools prefer to allocate whole afternoons to work of this kind rather than whole mornings which in most schools are rather longer.

Blocking time for team-teaching in this way makes the task of drawing up a timetable for the school rather easier than it is when one has to time-table single and separate periods for each subject. In the long term, there-fore, it is to the administrator's advantage also to make this kind of provision. In the short term, where such an arrangement is to be super-imposed on an existing timetable of a traditional kind, it can be a real headache. What one must try to do is to run together the time allotted to those subjects that have opted to 'come in' on the team-teaching pro-gramme. Another solution is to switch to double-period timetabling for all purposes, so that one is looking for blocks of two double periods rather than three or four single periods to run together for team-teaching and one can in any case, if in difficulty, offer double periods by themselves, since this now gives eighty minutes rather than forty.

The other main area in which team-teaching is dependent on the efforts of the administrator is that of accommodation. The two central con-siderations here are that there should be a variety of rooms available, or that what rooms are available should be capable of a variety of uses, and that they should be as close to each other as possible. It has already been pointed out that one of the advantages of team-teaching is that the size of teaching units can be varied to suit all possible educational needs. This can only be done if accommodation is available to suit teaching groups of all sizes, including a room that will accommodate the whole group at once. Some new schools have purpose-built accommodation for this kind of work with large rooms that can be divided into several small rooms by partitions or that can be given a number of 'corners' where different kinds of activity can be centred. However, few schools have the advantages of flexible spaces of this kind and in practice most have to rely on the hall and several adjacent classrooms (another time-tabling complication when the hall has to be used as a gymnasium and perhaps as a dining-room also). It is interesting to note that some of the older buildings with several rooms opening onto a central hall offer better accommodation for this kind of work than many of the modern palaces of glass and concrete.

Rooms need also to be of different types as well as of different sizes. Working from children's interests can generate a lot of practical activity — dance and drama, as well as graphics and model-making — and workshop facilities of all kinds are needed. Where such facilities cannot be provided on a permanent basis, it is sometimes enough to provide occasional help by allocating such accommodation for part of the time allotted to team-teaching or by arranging for the occasional pupil to work in a corner of a workshop if he has something urgent to be done — handicraft teachers are usually particularly ready and able to allow informal work of this kind. However, if such facilities cannot be made available, one can do no more than discourage children who express a desire to do something that requires them.

The last important requirement in accommodation for team-teaching is that the rooms allocated are as close to each other as possible. No team should be expected to play at home and away at the same time. I have known situations where the rooms allocated to the team were at different ends of a large building. Not surprisingly, in such circum-stances no real team-work can flourish. Movement between groups ought to be as easy as possible and there ought to be ample opportunity for teachers and pupils to see the work of the other groups as well as of their own. Without such minimal interchange, there can be little that merits the name of team-work and, while formal arrangements can be made to allow for it, there is much to be said for informal and continuous interchange. In any case administration ought to make the teacher's job easier not more difficult.

Planning and organization

Given an administrative set-up that is as supportive as it can be in the particular circumstances of the school, teachers can begin to plan their work as a team. As in Chapter 3, it will be assumed that if it is the intention to work through an agreed and preplanned syllabus, teachers will need no help with how to do that and attention will be concentrated on what is needed for the planning and carrying out of an interest-based, enquiry-based project. The general idea will be to present an area of work to the pupils in the most stimulating way possible and to encourage and assist individual and group enquiry within that broad area. This approach was discussed at some length in Chapter 3 in relation to the work of individual teachers. We must now consider some particular points that need to be borne in mind when it is being carried out by a team of teachers.

Paradoxically, the more freedom it is intended that the pupils should have, the more detailed needs to be the preplanning and organization of the project. This is the first point that must be made clear. Many teachers feel that there can only be freedom for pupils if they go into their classrooms without any plan or preconceived ideas, to 'see what happens', 'play it by ear', 'allow it to develop spontaneously' or with some other such intention that makes them sound more like jazz musicians going to a jam-session than responsible educators setting about their task. The best one can say about this view is that it reveals some confusion over the concept of 'freedom'. The only freedom a teacher can be concerned with is that kind of freedom that is not at odds with the requirements of an educational situation. To create an educational situation requires great skill on the part of any teacher and, no matter how experienced he is, a great deal of forethought, preplanning and careful organization. This is especially necessary when it is a coordinated piece of team-work that is required. Nothing that can be done in advance should be left to the hurly-burly of the classroom itself, but at the same time one must not preplan that which should be left for on the spot decision and thus impose a rigidity that will deprive both teachers and pupils of the kind of freedom this approach is designed to give them.

The first point that will need to be clarified is the area of responsibility of each member of the team. Some teams will include administrative staff; in time there may be assistants or aides formally employed in schools, some of whom will be attached to teaching teams; some teams may already include other nonteachers invited to make some specific contribution to the work. It is assumed that the role of such members of the team will be clear.

The position of the teacher-members must be made equally clear. Some, or all, may be expected to take existing forms or classes within the project, particularly where, as perhaps with younger pupils, it is intended to maintain the security that comes from membership of a single, stable group, or it may be the intention to regroup the pupils according to the areas of their interests, so that some or all of the teacher-members of the team may be required to take responsibility for groups of pupils working in particular areas. A team preparing for the project on 'Communications' referred to in Chapter 3, for example, would clearly be well advised to plan for groupings of pupils centred on the historian, geographer, scientist, social scientist, English and expressive arts specialists. Pairing of teachers might also be used where appropriate, two teachers being given joint responsibility for a larger group. On the other hand, it may be the intention that one or two teachers be given a roving commission — to help with the graphic presentation of the work of pupils in all the groups, for example, to guide all the practical work that is done, to attend to the audio-visual needs of all groups or to provide any service of this kind that makes it impossible or undesirable that they be limited to one group. It may be possible, of course, to arrange that both functions be fulfilled by one person. There are many possibilities here. What matters is not so much the particular form of organization that is chosen as that each member should be clear about it and about his own responsibilities within it and that at any one time every pupil should know which teacher is immediately responsible for his work.

It is also very important that the method of decision-making is clearly specified and this raises the question of the leadership of the team.

In many schools, leaders of teams have been appointed by the head-teacher; in some cases such appointments have been formalized by the award of some kind of responsibility allowance. There are difficulties in this kind of arrangement. In the first place, it has often worked out that such a leader, while responsible for the work of the team, has not been himself an active member of it, since his administrative responsibilities elsewhere in the school have made it impossible for him to involve himself fully, or at all, in the team's project. Clearly, decisions should not be made by someone who, no matter how competent and experienced a teacher, has not an intimate knowledge of what is going on. However, even when such an appointed leader is an active and full-time member of the team, there are difficulties in the notion of collaborative team-teaching in a situation where the team is to work under the direction, however benign, of an appointed leader.

On the other hand, similar, if not more serious, problems arise if one allows a leader to emerge from the membership or to be elected from it. Such a person can be more dominating and restrictive of the freedom of other members of the team, albeit through charm, personality and ability rather than official status, than any appointed coordinator.

On the whole, it would seem better to maintain the democratic ethos of the team-teaching approach by adopting some form of rotating chairmanship, allowing decisions to be taken in an entirely democratic manner and requiring each member of the team to take his turn as chairman having executive responsibility only and being a general factotum and dog's-body for about a week at a time.

Sometimes a team project may be based on a 'core' subject, or a 'core' subject may appear as the work proceeds. In this kind of situation the teacher responsible for that subject will inevitably play a leading role in planning and carrying out the project. Where a 'natural' leader of this kind emerges, there is little point in depriving him of the opportunity to guide the work of the team, since he is best qualified to do so. Furthermore, it sometimes happens that different subjects and specialisms come to the fore at different stages in the development of the work. To allow the teachers associated with these subjects to take on the leadership of the team when their own specialism is in the ascendant is to allow a kind of natural rotation that would seem to offer the best of both worlds. Again, however, the really important thing is that the method of arriving at decisions should be clear to all from the outset.

This is, of course, one aspect of the larger problem of the extent to which teachers should be able to participate in the making of decisions within the school. The arguments just offered would seem to have a similar application here, since, as we said earlier, organizational and other changes will only work if teachers accept them and believe in them and teachers can only work successfully in the way we are suggesting if they have some genuine say in the organization of their work. If team-teaching implies collaboration, as it must, then this must be taken to its logical conclusion as one aspect of a wider pattern of collaboration within the school.

Once the team has been given a definite structure in this way, attention must be turned to planning for the work that is to be undertaken with the pupils. In the first place, a clear view of the principles or aims of the project should be reached by discussion, which it is hoped will end in some kind of mutual agreement. Then attention must be turned to method. There are two aspects of this. First, as much detailed preparation as possible needs to be made for the activities that pupils are likely to be engaged in and the subject-matter they are likely to need to have presented to them in one form or another. Secondly, careful planning is necessary of an organizational structure that will ensure the continued smooth running of the exercise.

Preparation of material for pupil enquiries and other forms of pupil activity can be broken down into the three areas discussed in Chapter 3 — selection of a theme, planning an 'impact' session or some other form of initial presentation and preparing for the demands that will be made by the pupils if an impact is achieved. Much the same considerations apply when this approach is adopted by a team as when it is undertaken by an individual. The theme should be chosen and its presentation planned with due regard for the interests and the areas of competence of all members of the team, such interests and areas of competence being thought of in wider terms than subject specialisms. As was mentioned before, those areas should be emphasized that are less likely to occur naturally to the pupils themselves, those aspects of the theme that the overall aims of the project or its basic principles would indicate need to be emphasized and those areas that it is felt might be less popular.

One must also have due regard for organization and avoid the embarrass-ment of finding three-quarters of the group choosing one area, with the rest divided among the other three or four areas prepared for. Some balance is necessary and we must work for it from the beginning. If we are promoting a study of the local environment, for example, we can take it for granted that some pupils will want to undertake a traffic census, or explore the history of the local football team or unravel the mysteries of the local bus service without any stimulation from us. Few will immediately think of looking into the pollution problems of local industry, the leisure problems of local youth or the cultural opportunities the neighbourhood does or does not offer. Yet these may be the very areas of enquiry our team is most competent to help with and they may be felt to be of more educational value to the pupils than those they naturally come to themselves. These, then, would be the areas to stress in our initial presentation — by visits, films, talks from local people and so on.

We must next prepare for the fevered **enquiries** we are hoping to provoke. We must be able to get those who are ready down to work immediately, before their enthusiasm wears off, and to do this we must have the necessary materials and apparatus prepared and some clear indication of how they can best get started. Work-cards, provided we take care not to make them too restrictive, can be very helpful in getting groups of pupils and individuals started and we can prepare this sort of thing for as many lines of enquiry as we have been able to envisage in advance. We must also do as much work as possible ourselves in the area or areas of enquiry we are to be responsible for, since we need to be able to deal with questions that the pupils will raise, we need to be able to offer many ideas for lines of enquiry to pupils who have not been initially fired to explore anything and we need the background knowledge that will enable us to lead all pupils on from where they start to something that is worthwhile and which stretches them. All of this preparation needs to be done by the team and for the team; teachers must learn to live with a situation in which material they themselves prepare is used by others. This is what collaborative team-work implies.

No matter how thorough the preparation of material has been, little will be achieved unless equal care is taken in creating an efficient organizational structure within which the work can go on. A programme must be prepared of any further sessions that are planned with the whole group or any other large unit together. Everyone in the team must know if and when such events are to take place and everyone must be carefully briefed by those responsible for such occasions about their purposes and the ways in which they can be followed up. Very detailed arrangements must be made for the initial grouping of pupils if this is to be done on a basis other than the existing class divisions, and for their movement from one group to another if and when this becomes appropriate; when, for example, a particular line of enquiry has ended or has led on to further explorations that can best be guided by another member of the team. Arrangements should also be made for pupils to see each other's work. This is especially important where each is being allowed to explore in depth one aspect of the theme only. Although there is a lot to be said for permitting informal sharing of findings, if each pupil is to see other aspects of it, there must also be formal arrangements which will ensure that sharing does go on — regular reporting-back sessions, ongoing exhibitions or something of this kind.

Perhaps the most important area where detailed organization is required is that of keeping an up-to-date record of the whereabouts, the assignment and the progress of each individual pupil. Unless this can be done, team-teaching will offer countless opportunities for time-wasting, repetition and even truancy. A careful record must be kept of all the children for whom each member of the team is responsible, so that none can go 'absent' without being detected. Although interchange of pupils between groups is to be encouraged, actual movement about the school during lesson time is in most cases not advisable. If it is felt that a particular pupil should be moved to another classroom, arrangements should be made between the teachers concerned in advance of the lesson so that his movement can be checked at both ends. If there has to be informal movement, it is better that it be of staff than of pupils. It is vital too that a detailed record be kept of what each pupil does, partly to ensure that he is doing something and that what he is doing represents a fair amount of work for that particular individual, and partly to see that he is not engaged yet again on the same piece of work that he did in last term's project and that of the term before. It is only detailed organization that will protect the teacher from wastrels and enable him to ensure that all pupils get a balanced educational diet.

Finally, the winding up of an exercise of this kind needs to be carefully planned, as does a method of evaluating what has been achieved. Again, it is essential that the team plan this as a team, so that the way of drawing things to an end can be agreed by all and the methods of evaluation such as to ensure that what each member of the team has been doing can be properly assessed. Almost certainly it will not be possible to make plans for this sort of thing before the work begins; in fact, it is most unlikely that it will be possible to do much about it until the project is well under way.

This underlines the need, therefore, for continuous planning and evaluation throughout the period of the project. A great deal of pre-planning must be done but this will achieve little without regular and continuing opportunities for consultation, evaluation, recording and further planning. Preplanning can provide too rigid a structure; this can only be avoided if the team meets regularly to evaluate progress and make further plans in the light of experience. It cannot adequately cope with the problem of wastrels and other difficult pupils; they can only be attended to if the team meets regularly to discuss, to share experiences and to keep detailed records of what each group and each individual has done. Often this is left to informal contact of team members over coffee, lunch and so on. Too much is at stake for it to be left to casual contact of this kind. Team meetings should be formally timetabled to take place regularly when all members of the team can attend. They should be as frequent as seems necessary, but there should never be fewer than one per week. The success of team-teaching depends on careful organization and the coordination of the efforts of all members of the team. This can only be achieved by ensuring that all planning is done formally and jointly and that each member has as full a view as possible of the general context in which he is working.

Some of the crucial issues that arise for discussion at such meetings are described by Don Steels in his account of the meetings of the humanities team at Northcliffe Community High School (1975, pp. 121-122).

> The weekly team conference (which is time-tabled) is essential in maintaining and sustaining the course and it needs to be supported by extra untimetabled conferences. At these, planning of a preliminary nature takes place long before the work is to be commenced and is followed by more detailed discussion on the course. Are we to make any alterations in content or application? What problems do we foresee? We feel that it is vital to vary any presentation of material and any organisation in order that the method does not become stereotyped and thus lack the necessary vitality which is of paramount importance to the course. How are we to use the specialist knowledge of the staff? I give but a few of the issues which can arise in our meetings. Having then received as it were a directive from my team I, as leader, then proceed to produce a timetable in order to provide a structure within which further detailed planning can take place.

Given a structure of this kind with regular opportunities to acquire a developing understanding of the goals and general principles of the work in hand, the individual teacher must, in the light of that understanding, guide the work of the pupils he is particularly responsible for, leading them on, passing them on to other colleagues if and when this seems the right thing to do, ensuring that all are working, that all are stretched and that all are getting value from their work. All of this can be left to his professional judgement.

Problems

We must now turn, finally, to a consideration of some of the particular problems that arise when a team-teaching approach is adopted and some of the more common pitfalls to be avoided. All of the difficulties that we discussed in the last chapter in relation to the adoption of an assignment approach by the individual teacher will, of course, apply equally to the team-teaching situation. Undirected activity, a 'hobbies' approach, is as easy to slip into in a team context as it is when working on one's own. Similarly, it is just as easy to lose sight of the opportunities offered for creative work, except in so far as there may be a member of the team whose specialism is of this kind and who will or should make it his business to keep this to the forefront of things. The difficulties of certain subjects, especially those that require the learning of certain skills, also need as careful consideration by the team as we tried to show they merited from the individual teacher. All of these difficulties must be kept in mind and planned for, but all of them were fully discussed in Chapter 3 and, therefore, need no further elaboration. Team-teaching, however, brings further problems of its own and these must be considered here.

In the first place, we must be aware of the host of psychological problems that arise for teachers from the kind of reorientation that team-teaching requires of them. We have already referred in passing to the need for teachers to accept that work they have prepared might be used by other members of the team. This is only one aspect of what becomes necessary when teachers move from a situation in which they have had a great deal of personal and individual freedom to one in which they must modify all aspects of their work and accept on all fronts those limitations that become necessary when several people work together. A degree of personal autonomy is lost by all members of the team in decisions concerning the principles, goals, content and methods of their work. Each member must to a large extent accept the goals and the structure of the team, since unless he does so the team cannot function smoothly, it cannot work as a team and will quickly break up. A common philosophy must be developed which has, for the most part, the allegiance of all the teachers involved, since again no team can get very far if there is no agreement over principles and priorities. The development of such a common philosophy will usually entail a certain amount of compromise on the part of all the team's members and such compromise is not always easy to make.

At least two things follow from this. First, teachers must learn to gain their satisfaction from the successes of the team as a whole and not merely look to their own personal achievement. They will thus perhaps become reconciled to seeing others using material they have prepared, others operating in spheres they may perhaps have once thought were their own and others learning from them and copying their ideas. They must learn to cast their bread upon the waters, to accept sharing as a normal part of the job and to evaluate their work in terms of its total value for the pupils rather than the sense of personal achievement it gives to them.

Secondly, we must accept the fact that not all teachers may be temperamentally suited to team-teaching, not all may be able to make the psychological adjustments necessary for working in such a context. This will be especially the case with teachers who have not been prepared by their initial training for work of this kind and this may explain why many headteachers have found particular difficulties when older members of staff have been asked to involve themselves in this kind of work.

On the other hand, it may be the case that what really matters is not the temperament of the individual nor his age nor the type of training and experience he has had, so much as the relative temperaments and backgrounds of all the team members. It may be that difficulties experienced when older teachers were attached to teams arose not from the fact that they were older as from the fact that they were older than other members of the team and that the spread of age and experience made it difficult to develop the community of purpose and interest that we have said is so vital.

In practice, both kinds of situation will be found and both need to be guarded against. Some teachers of all ages will never be able to make the necessary adjustments to enter a team project; all teachers will only be able to do so if they can achieve a psychological 'fit' with their fellow team members. We must also guard against a situation that will inhibit or stifle the individuality of the brilliant but idiosyncratic teacher. Such a teacher has a great deal to offer if the team can adapt itself to his idiosyncracies. If a working relationship cannot be developed, however, it would be better to let him have his head in another context where collaboration is not so essential.

Some of the main features of this change to team-teaching as it affects individual teachers are again described by Don Steels in his account of the work of the humanities team at Northcliffe Community High School (1975, pp. 114-115).

> The teacher was a member of a team; he concerned himself with the development of the child; he was a consultant and friend, a director and

an authority, a ways and means man, and a person who was sought out by pupils and staff for advice. He was no longer in a little box in isolation, where he was the ultimate authority; he now had to change his role to fit in with the new philosophy. This was a gradual process — a process which was nowhere near complete. Many accepted and welcomed the situation; others perhaps were uncertain and continued to teach as they always had. This transference from being the planner, the instructor, the disciplinarian in his small domain, to the more demanding and stimulating test of the new situation, was difficult to make.

In addition to a reorientation to their work and the development of new attitudes towards and relationships with each other, team-teaching requires teachers to engage in a lot of rethinking of their own subject specialisms. Team-teaching necessitates a reappraisal of the content of our teaching in relation to other subjects, to other areas of the curriculum. As has been pointed out, team-teaching is possible within subject areas, but most team-teaching projects span a number of subjects and, although it is possible to retain subject boundaries within such a project, in practice these barriers become blurred, even if they do not disappear altogether; in fact, team-teaching is often undertaken with the specific intention of breaking them down. Furthermore, if team-teaching is also associated, as it would seem it should be, with an enquiry-based, interest-based approach to learning, it is difficult to retain such barriers; subject integration in some form will follow almost inevitably and this will involve a reappraisal of the role of each contributory subject.

It is not my intention here, nor would it be appropriate, to become involved in the arguments that rage over the respective merits of disciplinary and interdisciplinary teaching and learning. As so often, the important question is not which approach we should adopt to the exclusion of the other, but rather in what situations each is most appropriate. It is quite clear that the logic of a subject, the rigour of a discipline, can never be ignored in the teaching of it. It should be equally clear that there are many areas in which we need to develop the understanding of our pupils where an interdisciplinary approach is unavoidable. The work of the Schools Council Humanities Curriculum Project (Stenhouse 1968; Schools Council 1970) has drawn attention to many areas in which integration of this kind is vital. No adequate examination of racial problems or relations between the sexes, for example, can be undertaken within any one discipline; issues of this kind must involve us in excursions into many subject areas, if we are to come to grips with them successfully. Indeed, quite often an examination of topics of this kind will take pupils into 'subjects' other than those traditionally associated with the school curriculum. It is not possible, for example, to examine the problems of relations between the sexes without considering the anthropological evidence of the pubic and marriage rites of other cultures, the law and its pronouncements on divorce, prostitution, pornography, homosexuality and abortion and the psychological factors involved in sexual relationships, in addition to the biological, sociological, historical and literary aspects of this topic that are already touched on in a traditional curriculum.

Teachers must be clear, then, about what they are contributing to a team-teaching project. There is always a danger that they will want to evaluate it in terms of how much of history or science or some other discipline each pupil is involved in. These are not necessarily the appropriate criteria to employ. If the agreed purposes of the team are to develop the ability of the pupils to think historically, to think scientifically or to acquire the essentials of some other discipline, then this is the kind of criterion to appeal to, but if the goals are other than this — to develop a greater awareness of the local environment, an appreciation of man's achievements or an understanding of human relationships — then each teacher must accept that his 'subject' will become subsidiary to the stated end. Conversely, he may find he has other 'subjects' to contribute to the combined strength of the team, interests that hitherto his work has given him no real scope for. Again it is a matter of seeing the education of the individual pupil as the goal, rather than the propagation of a particular body of subject-matter or a particular discipline.

Finally, attention must be drawn to the fact that team-teaching does create particular problems for certain subjects. Mention was made in Chapter 3 of some of the difficulties experienced by teachers of foreign languages, mathematics and science in adapting to the individual approach required by the mixed-ability group. Similar problems arise when attempts are made to involve these subjects in a team-teaching scheme and again the solution lies in careful and rigorous analysis of the principles of the scheme.

More serious, however, to my mind are the problems that arise for art and craft specialists in a team-teaching situation. If they do not assert themselves, they can easily find themselves relegated to the role of advisers on the display and presentation of work that has been done in other subjects for other members of the team. Too often, art in a team-teaching project becomes a kind of handmaiden, concerned merely with illustration and presentation, and no really creative work is generated. This is easy to understand. For if one wants a diagram and there is an artist in the team, or if one wants a model made and the team contains a craftsman, who better to get to do the job? This is a role from which the art and craft teachers must be protected by the others and from which they must defend themselves. No teacher should be reduced to the role of teacher's aide nor should the opportunities which the presence of an art or craft teacher in a team can offer be squandered in this way. If the team is fortunate enough to contain an artist or a craftsman, their skills should be used to contribute to the education of the pupils just as much as the skills of anyone else.

Summary and conclusions

In this chapter we have discussed some of the advantages of adopting a system of team-teaching, noting the particular merits of such an approach to the teaching of mixed-ability classes and especially the opportunities it offers for that flexibility of grouping that it was suggested earlier should be the key feature of any mixed-ability form of organization. We then considered some of the administrative arrangements that need to be made to create a structure in which it can be used to greatest effect before considering the most important things that teachers need to keep in mind when planning and carrying out a collaborative programme of work. Finally, we looked at some of the difficulties that must be overcome if such a system is to be employed with full effect.

In spite of such difficulties, team-teaching offers teachers a great deal of scope for new, interesting and rewarding work. It is one way of taking advantage of some of the opportunities offered by a mixed-ability form of organization and an increasing number of schools are adopting it in association with this kind of reorganization.

Whether one sets out to tackle a mixed-ability class alone or in a team, however, there are many other facets of the task that have to be considered if success is to be achieved. We must now turn to a discussion of some of the more important of these.

CHAPTER 5

GROUPS AND GROUPING

Groups play a large and important part in the lives of all of us. We are all members of a number of small groups — the family, the office or department, our circle of friends at the pub or club or church — and our membership of these groups is perhaps the single most important factor in our lives. Certainly it is through membership of these groups that our lives and attitudes are shaped and modified, and it is to these groups that we owe our most easily recognizable loyalty. We find it easier to see ourselves as members of small groups of this kind than of society as a whole. This is what Aristotle meant by his often misinterpreted dictum that 'man is a political animal', since he had in mind the Greek view of 'political' as referring to the life of a 'polis' or small city community. There is much anthropological and indeed zoological evidence to support this view of man's gregarious nature.

Furthermore, there is a growing body of psychological evidence which suggests that the group, and especially the peer group, is particularly important at the adolescent stage of development for most young people — a fact that we can hardly doubt as we watch adolescents moving about our cities like clouds, ever changing their shape, forming and reforming, but seldom to be seen unaccompanied. Clearly, while we all draw strength from membership of groups, the adolescent seems to have most need of such strength, perhaps because for the most part he seems to be able to take it from only one source — the peer group. Psychologists have also suggested that there is a need for adolescents to develop a 'self-image'. They have further proposed that this will be the product of the interaction of the individual with others (Mead 1934; Lindesmith and Strauss 1949; Manis and Meltzer 1967). Amongst these 'others' the peer group clearly plays a very important part.

If we accept that it is the duty of the school to support adolescents through this period, to help them with this task as well as to attend to their academic and intellectual needs (and it could be argued that the move to mixed-ability grouping in itself implies an acceptance of this duty), then we must keep this dimension of our work in mind in all our planning. It might be felt by some that it is enough to include specific elements in the curriculum to attend to this aspect of the development of our pupils — special lessons in moral education, for example — and to provide adequate supportive counselling services. But the message of the psychologists is clear; we develop our self-image through all kinds of interaction with others, not just in those situations artificially contrived towards that end, so that an awareness of this responsibility should permeate all our dealings with adolescent pupils and indeed, since this process does not start from scratch with the onset of puberty nor end at its completion, with pupils of all ages.

A second and related reason why attention needs to be given to the possibilities of group work is that the change to mixed-ability classes, as we suggested in Chapter 1, represents a fundamental change of educational thinking, one feature of which is an increased concern with the development of social relationships in our classrooms, a shift from a competitive to a cooperative or collaborative view of education. If this is so, then there is implied the need for us to consider how much of the work of our pupils should be undertaken on a cooperative basis and what form this should take. It is through his work in school as much as from any other source that each pupil will learn to come to terms with himself as an individual in a social setting.

For both of these interrelated reasons we need to explore more fully than we have hitherto the possibilities of getting our pupils into suitable groups and to seek for devices we may use to promote work of a collaborative kind. The individual assignment has its place but if work becomes too individualized the social advantages that are claimed for the mixed-ability class are lost. We need to consider very carefully, therefore, how we can encourage group working as well as linking the work of individuals in the ways we have suggested in the last two chapters.

Criteria for grouping

First of all, some attention needs to be given to the basic criteria upon which a teacher might subdivide a class into groups. There would seem to be at least four broad approaches the teacher could adopt, all of which one can find being practised at present.

First, there are many teachers who, faced with a mixed-ability class, will group the pupils according to their abilities; in other words, they will solve the problems presented to them by the unstreamed school by streaming within the class (Barker-Lunn 1970). Pupils will be grouped according to their previous achievements in the subject concerned and can then be pushed on at a rate that is right for their level of attainment in a group with others whose pace of working is roughly similar. In this kind of situation the teacher prepares not one class lesson but four or five group lessons, and thus covers the content of the course at a rate appropriate to each group.

The advantage of such an arrangement is that it provides the teacher with relatively homogeneous groups to work with, while avoiding some, if not all, of the social difficulties associated with streaming on a large scale. Furthermore, it is relatively easy for such groupings to be changed for different subjects or to allow for the different rates of progress of individual children. Thus it would be claimed that a pupil can be working with fellow pupils of roughly equivalent ability in all subjects where this kind of approach is felt to be relevant and can be readily moved 'up' or 'down' as his position changes in relation to the others. Only thus, it would be argued, can we ensure that all pupils will be stretched and none broken. This is possibly the most common form of grouping that one can find in schools at the present moment and a good deal of so-called 'unstreamed' teaching in primary schools is of this kind.

On the other hand, it is claimed by some that grouping by ability within a class suffers from all, or most, of the difficulties associated with grouping by ability within the school. In particular, extraneous factors, such as social and personal characteristics, enter into the teacher's judgement, and the groupings that emerge turn out not to be based solely on ability at all. Furthermore, the streamed grouping within the class then gives rise to the kind of 'self-fulfilling prophecy' that has been shown to be associated with streaming within the school (Jackson 1964), and pupils begin to work at the level of the group they find themselves in, to satisfy but never to exceed the expectations their teachers have of them. Even though such groupings appear to be based on ability, therefore, many able pupils are not stretched, since for other reasons they have not been placed in a group whose expected level of work makes real demands of them.

Many would want to argue further that to adopt this kind of approach is in any case to miss the point of unstreaming and the opportunities it offers. Such teachers would want to use something other than ability as the basic criterion of grouping. An extreme form of this attempt to get away from academic ability as the only essential criterion for organization within a class would be to group pupils in an entirely random fashion — to draw names from a hat or arbitrarily send them to different tables or corners of the room as they come in through the door. Such an approach might be likely to ensure mixed-ability groups within the mixed-ability class, but could lead to some injudicious combinations, and would equally ensure that the teacher gained none of the advantages that are to be gained from other forms of grouping. It is not, therefore, to be recommended to any except confirmed addicts of Russian roulette.

The third basis one might have for grouping is a modified version of this approach, and would certainly be favoured by many teachers — the grouping of pupils according to friendship patterns within the class or, more strictly speaking, allowing pupils to group themselves according to friendship patterns. Friendship is a more important factor in school than perhaps many teachers realize. In particular, it has an important bearing on behaviour. It is unwise for teachers to ignore such friendship patterns, therefore, but whether they should go to the other extreme and use them as the only basis for grouping pupils is questionable. In the main one would want to question it on the grounds that it places the entire responsibility for grouping in the hands of the pupils themselves, and although the wishes and choices of pupils are important and highly relevant factors in such decisions, there must be some teacher direction too. Teachers must take control to ensure that the groups that are formed are the most appropriate for the work in hand. For here, as in all aspects of educational decision-making, it is the teacher's job to use what advantages a situation offers to further his pupils' education.

A basis for compromise between pupil choice and teacher direction is offered by a fourth method of grouping, grouping by interest. This method presupposes, of course, an approach to teaching like that discussed in Chapters 3 and 4, namely an interest-based, and perhaps also an enquiry-based, approach. If this is to be our approach, then it clearly makes sense that children should be grouped according to the interests they show, and they must be given time to get to know each other's interests and to group themselves accordingly.

It is interesting to note the overlap here with grouping by friendship, since there is some evidence to suggest (Allen 1971), and many teachers will themselves have observed, that the interests pupils have outside the school seem to prove stronger than those developed in school and that friendship plays an important part in choice of activity and interest in any kind of project. This is in itself, of course, a reflection on the importance of friendship to which we have already referred and an argument for providing within the school opportunities for such relationships to develop.

Some teacher direction is, however, also necessary. It has already been stressed more than once that teachers should use their pupils' interests to promote their education, and this process begins at the point where groups are formed. Even if interest, along with friendship, is the basis, teachers will need to make adjustments, and such adjustments should be made in the light of what is known about relationships within groups and of the purposes and principles behind the adoption of group methods.

It must also be stressed that all groupings must be flexible, the groups must change form and reform according to the changing purposes and contexts of the learning that is going on. No one grouping of pupils into subgroups within the class can meet all the purposes of the teacher any more than any one kind of grouping of pupils into classes themselves. Again we note both the need and the possibilities for flexibility.

Academic and social aspects of grouping

There are two main reasons why teachers will want to divide pupils into groups and, correspondingly, two main purposes that should be borne in mind when deciding on the methods by which they are to be so divided. Clearly, one purpose will be that of furthering their academic education.

Group-work can be viewed as a methodological device for providing each child with the appropriate educational diet. Seen in this way, its main purpose will be to enable the teacher to ensure that all pupils are fully stretched and that none are asked to tackle problems that are beyond them. If this were the only purpose of group-work, then it might be thought unnecessary to look beyond the first method of grouping discussed above, namely grouping by ability. However, implicit in the criticisms of this method is the notion that the furthering of the academic education and scholastic attainment of each pupil is not the sole aim of grouping. A further aim, which is embodied in the idea of grouping according to friendship and interests, as we saw at the beginning of this chapter, is that of furthering their social and personal education and this is the second kind of consideration that teachers must bear in mind when forming groups.

Some teachers, of course, would want to deny that social learning or the personal develpment of their pupils was their concern and would wish to assert that their only purpose is to further the academic progress of their pupils. This position would seem to be reinforced by recent moves towards establishing 'moral education' or 'social education' as separate 'subjects' on the timetable, developments which suggest that they can be left to the teacher responsible for those periods.

It is doubtful, however, whether such a position is tenable. Time for looking at moral and social issues in a responsible way, under the guidance of someone skilled to supervise it, is no doubt of great value, but a major part of social learning and growth as a person is the development of social attitudes that comes about through real relationships, teacher-pupil relationships no less than any others. Social attitudes are caught rather than taught, a truism stressed in the assertion of the Newsom Report that teachers can only escape from their influence over the moral and spiritual development of their pupils by closing their schools. Social learning is the result of every teacher's relationship with his class and his method of organizing his work with that class. Indeed, there is evidence to suggest that a teacher's ability contributes relatively little to the academic success of his pupils in comparison with the marked effect it has on other aspects of their development, including their social development (Evans 1962).

Social learning and personal development, then, are functions of the social organization of the school and of the class, so that they are the responsibility of every teacher. They should not be allowed to happen in a random fashion and consideration should always be given to the likely results of one's teaching in terms of the social learning it is likely to promote. It is a puzzling feature of present-day educational debate that those teachers and others who are concerned lest new methods bring a random element into children's academic learning are often quite content to accept an even greater random element in their social and moral learning.

What kinds of social aim might one have in grouping pupils? What kinds of social learning might one want to encourage? At one level, one might merely wish to create a situation in which personal relationships between pupils can develop or in which pupils can learn to make personal relationships and at the same time come to terms with themselves and develop their own self-image. Mention has already been made of the importance of groups and personal relationships within groups to all of us. If this is so, then clearly teachers should do whatever they can to enable their pupils to learn to form such relationships and to get full benefit from them. Mention has also been made of the important part that friendships play in the lives of children and adolescents. Teachers should not undervalue such friendships nor should they lose sight of the opportunities such friendships present to them for fostering both the social and the academic development of their pupils.

On the other hand, an important aspect of the social development of pupils is that they should learn to respect and work with whomever they are required to work with by the exigencies of a particular situation or by someone who has a wider perspective on the situation than their own. Furthermore, it is only in this way that we can help them to acquire the ability to make new relationships and fit into different social contexts. The possibilities here are well illustrated in Shirley Legon's account of the methods of grouping she used with one class of girls in teaching sociology at Fairlop School (1975, p. 142).

> During this course of group dynamics, I kept changing the size and composition of the group, so that they were forced to work with people they did not know very well, although they had been in the same class for two years. One very shy, withdrawn girl told me later that as a result of this experience she had summoned up the confidence to join a Youth Club, since she had to stop relying on her one friend, also shy, for company.

The development of personal relationships of all kinds and the promotion of the pupil's ability to form such relationships must be among the basic social aims of grouping children. Many teachers, however, would want to go further than this and aim at the development of particular kinds of personal relationship. There has been much discussion in recent years of the respective merits of cooperation and competition in education. Some have argued that in a competitive society schools should encourage competition to prepare children to take their place in that society, and have looked for support also to the motivational advantages that a competitive situation seems to offer. Others have offered the counter-argument that it is part of the job of education to change society and not merely to fit children to it, that at least teachers should not compromise their professional commitment to the moral upbringing of their pupils, whatever kind of ethos exists in society, and that an education motivated by competition is perhaps not strictly deserving of the name of education at all.

It is not the intention here to become involved in this debate, nor is it necessary, since whatever side one takes in this argument, there is no denying, as we have already suggested, that the change from streaming to unstreaming represents a move from a competitive to a cooperative system. It is the result of a general move towards a collaborative approach to education at all levels. In short, cooperation and collaboration are the fundamental principles of a mixed-ability class organization, so that in the context of our present discussion it must be taken as given that a further social aim of any subgrouping of a mixed-ability class will be the promotion of a cooperative and collaborative approach to work. The teacher should be trying to promote cooperation and collaboration and to discourage competition.

Therefore teachers must approach the task of grouping their pupils with both academic and social purposes in mind. Of course, these aims are distinguishable only at the conceptual level. In practice, they are closely interwoven, since it is clear that the social climate the teacher creates is a very significant factor in the academic development of his pupils and that many social relationships will develop out of work commitments jointly undertaken. Working with others, especially with one's friends, is obviously more attractive to many than the loneliness of working in isolation; it can add a new dimension to work the pupil enjoys and reduce the level of boredom when the work is routine. In fact much evidence suggests that the social situations in which pupils learn are as important as their intellectual abilities in deciding their levels of academic achievement (Amaria and Leith 1969). If this is so, the social environment must be carefully structured by the teacher to give himself the best chance of promoting his pupils' academic attainment.

Conversely, many of the social relationships we are concerned to develop will grow out of the academic work undertaken by the group. It has been argued that work is a group activity and that the social world of the adult is based primarily on his work activity (Brown 1954). This provides us with another argument in favour of taking seriously the social education of pupils and again, if this is so, teachers must look to the work itself as a source of social learning. In practice, then, academic and social goals will be inextricably interwoven and to concentrate on one to the exclusion of the other will be to risk missing out on both.

The formation of groups

To take full advantage of what can be achieved through grouping techniques in both of these spheres requires a knowledge of the ways in which groups form, some of their characteristics and the kind of inter-action that takes place between individuals within them. We must now turn, therefore, to a consideration of some of the factors that teachers should bear in mind when supervising the forming of groups within their classes.

Quite a lot is known about the ways in which children and young people will group themselves when given the opportunity to do so and the teacher needs to be familiar with these basic trends so that he can avoid the problems that will arise if he goes against them, and can take advantage of the opportunities that accrue if he understands and uses them.

To begin with, we have already referred several times to the effect of friendship patterns and out-of-school relationships and interests on the grouping of children in schools. Pupils, if given a free choice, will tend to choose to work with those with whom they have some contact outside the school, those who live near them, went to the same primary school or meet them at Church, Sunday school, Scouts, Guides or the Youth Club. This tendency is completely natural. We do it ourselves. When entering a situation for the first time we look around for people with whom we have had previous associations in other contexts. A glance at the groupings which form at any PTA dance, for example, will quickly reveal to us the truth of this.

We should also note that these tendencies will vary in form with the ages of the pupils concerned and that the stage of development will be a significant factor in the formation of groups. In the preadolescent stage, for example, boys and girls are unlikely to mix but will naturally group themselves separately.

We have already referred to the advantages that can accrue when pupils are able to work with their friends in social environments they have largely been able to create for themselves. They seem to get on better, faster and with less stress, so that there are fewer behaviour problems to distract the teacher from his main task. On the whole, then, there seems to be much in favour of going along with this general tendency and allowing these friendship patterns to determine largely the composition of the working groups within the class.

There are, however, some pitfalls to be avoided here which draw our attention again to the need for the teacher to exercise some control over the forming of groups.

In the first place, this tendency to maintain out-of-school relationships in groupings within the school might result in the concentration of all potential troublemakers in one or two groups. The common interest uniting such groups might be a shared desire to drive the teacher up the wall and this is not the sort of interest that provides him with scope for development into educationally profitable channels. The cement that holds such a group together may be its collective resentment of authority and this could well be the result of certain kinds of out-of-school association. After all, what distinguishes a group from a gang is no more than the different view taken of the purposes of each by society as a whole. Keir Hardie led a group; Al Capone a gang. In both cases they were at the head of small bodies of people united by a common interest and certain shared goals. In dividing his class into groups, the teacher must, therefore, avoid creating gangs or allowing such to develop.

It may well be the case, of course (indeed, one would hope that it would be), that this new approach to teaching will remove some of the causes of troublemaking and harness the energies of such pupils to more purposive and constructive activities. Certainly, by taking away the sense of rejection suffered by some pupils it should have removed what seems to be one of the prime causes of the generation of delinquescent subgroups (Hargreaves 1967, 1972; Lacey 1970). However, some will perhaps always remain and in the teacher's own interests, in those of the class as a whole and, not least, to the ultimate advantage of the trouble-makers themselves, he must distribute these pupils widely and judiciously among the other groups which are forming.

In doing so, he must remember that it is not enough merely to place such pupils physically at the same table or in the same corner as the group he wishes them to work with. To be successful a group must have real cohesion; all members must feel themselves a part of it and must be accepted by their colleagues as a part of it. In all cases where the teacher is adding members to groups, therefore, he must be aware of the need to show all concerned that there is a positive purpose in the change he is making, to show the group that the new member will have something of value to offer it and to show the new member that he has a role to play and a contribution to make to the collaborative enterprise.

It will also be an advantage if any adjustments of this kind can be made at a very early stage, since once a group has formed and begun to develop a sense of unity and group identity, it is very difficult for a newcomer to break into it.

A second danger that is endemic in giving a class complete freedom to subdivide itself is that there may well be a tendency for groupings to be based more on relative ability than the teacher may feel is desirable. Community of interests may well mean similarity of intellectual capacity, as might family or neighbourhood relationships or involve-ment in and attendance at the same out-of-school activities. If the teacher is aware of the importance of the social learning he is trying to promote, he will want to avoid this kind of development on any large scale and will want to ensure some kind of 'mix'.

Normally, in the nature of things, this problem will resolve itself into a matter of attaching the less able, the slow learners, the nonreaders to groups. Again, as in all cases where he is adding individuals to existing groupings, he will need to show all concerned that there is a valuable contribution to be made by such pupils. More will be said in a later chapter about the problems of children who have learning difficulties in the mixed-ability class. It will be enough if one merely says here that they should not be allowed to group themselves together and struggle along on some activity or project that can be seen by them and all others to be inferior. This cannot result in the learning, social or academic, that the teacher would want for any of his pupils. Nor should they be separately grouped to have 'remedial work', while the others are pursuing their enquiries and their interests. It should not be difficult for a teacher to discover some distinctive contribution that such pupils can make to a collaborative undertaking. There are many pupils whose reading ability is poor but whose skill with paint and crayons and general talent as illustrators or whose facility in handling electrical equipment or setting up other apparatus is of a high order. If social learning is to take place, we must show all pupils that educational enterprises do not always involve the skills of reading and writing, that other kinds of activity can be just as valuable, in some contexts more valuable, and that we do value these things. So much that is wrong with present-day education, especially that of the adolescent, springs from the tendency to disvalue the skills and interests of young people and create a gap between their values and those of the system. The mixed-ability class must be seen and seized as an opportunity to bridge that gap.

The third main concern of the teacher as he supervises the formation of groups within his mixed-ability class must be the situation of the isolate, the pupil who, for whatever reason, has no friends and is not generally acceptable to the existing groups. Such a pupil will, like all others, have his own interests, will choose a group to work with and has an entitlement to be allowed to work with it. Every pupil must have the opportunity to gain the support of working with others if he wants it.

The teacher's task here is particularly difficult. Not only must he be able to show a group that the new member he is offering it or who wishes to join it has a valid contribution to make to its work, he must also overcome the objections to working with this pupil as a person that other pupils will raise in the light of their previous knowledge of his unattractiveness. Furthermore, the teacher has in this situation a remedial task to perform with the individual concerned, for such a pupil is clearly backward in his skill at making personal relationships and part of the teacher's job in getting him involved in a group is to give him the help he needs in making friends and developing relationships with others. The teacher will want to avoid, therefore, the situation in which he takes such a pupil along to a group to be met by a chorus of voices loudly declaring their unwillingness to be lumbered, particularly as the unfortunate pupil concerned may have already had this experience before taking his problem to the teacher.

In this kind of situation, it is especially important to move in before groups have become too fixed or have achieved too strong a sense of group identity. The teacher needs to know in advance, therefore, who the isolates are and which pupils will need particular attention from him in the early stages of group formation. He would also be well advised to keep all groups in a reasonably loose and fluid state until all of the adjustments he feels are needed have been made.

In concerning himself with this difficult problem of the isolates however, the teacher must avoid confusing them with the solitary workers. There are pupils who choose to work alone for reasons other than an inability to make relationships or to work with others. They may not choose to work alone on all occasions, but sometimes they like to. In some cases this is because of a feeling some pupils seem to have that less work is done in a group project, that they will 'get on' less well and that to ensure progress they should work by themselves.

It is difficult to generalize about such pupils. Certainly one would be reluctant to force them into group work and forbid them to work by themselves. Indeed, it may be important that some pupils should learn to work alone, since many jobs require this, not least those involving study. On the other hand, it is clear that when they do work alone there can be little gain in their social education and if one of our purposes in adopting group-work techniques is to assist the development of their abilities to work with others then we must be aware that we are not succeeding in this with the solitary worker.

The answer here would seem to be to allow individual working, since one does not want to cut too violently across what may be a well-developed learning style, but to keep a close watch on and careful record of those pupils who take advantage of it. It is really only those few who always choose to work alone that the teacher need concern himself with. Something can be done too by encouraging all such pupils to join in with others for occasional short-term and subsidiary projects. Flexibility of groupings is as important here as elsewhere.

This question leads us on to the general issue of the optimum size of groups. In addition to the solitary worker, the teacher will observe also the phenomenon of pairs. This kind of paired grouping is a feature of certain age-groups and seems to be more common among girls than among boys. Again it is something for the teacher to keep an eye on rather than to get anxious about and again it can best be dealt with if the groupings are flexible and if for some purposes such pairs can be integrated with others.

At the other extreme, there is the question of how large a group should be allowed to become. Clearly, the answer to this question will vary according to the purposes of the group concerned. It should be large enough to achieve its purposes but not so large as to exclude some members from making a full contribution to its work. Full participation of all members can probably be achieved if the group has four, five or six members. It is also in such small groups that social learning can best take place so that four to six would seem to be the norm to aim for. In practice, it is unlikely that most groupings will attain a membership above eight. When they begin to reach that size, there is a natural tendency for them to subdivide themselves. This in itself is not a bad thing, since it creates some interaction between groups and this is difficult for the teacher to generate from the outside.

In supervising the formation of groups within his mixed-ability class, then, the teacher must keep a close watch on the progress of the trouble-makers, the slow learners and the isolates and must attend to the size of the groups that are forming. He must also bear in mind the need to match the learning styles of individuals and their personalities or at least to avoid the possibilities for clashes. Little is known at present about the effects of different personality characteristics on learning as such or on group learning, but individual personality will clearly have a great effect on the social climate within any group and the relation-ships existing between its individual members. Trouble for the teacher does not always arise from confirmed troublemakers; it can come about when two normally well-behaved pupils are set to work together and evident personality clashes result. Teachers need to give attention to this factor too, therefore, in the formation of groups.

Such clashes are especially prone to arise over the leadership of groups. A particularly strong and dominant character will tend to endeavour to assert authority over the group as a whole and two such characters in the same group could be a source of trouble. In this connection it should be noted that it is not always the most popular pupil who comes to leader-ship in such situations; more often it is the one who has the most ideas and is able to come up with suggestions which all can see are the best suggestions they have had for forwarding their common purposes.

However, we might also ask whether teachers should encourage the emergence of group leaders at all, since it would seem that, while it might do a lot for the pupil leader, this might be achieved at the expense of the development of qualities of self-reliance and confidence in the rest. Furthermore, it may not advance the learning of all to be in a situation where one pupil assumes control, since the others may come to feel alienated from the purposes of the group and motivation will thus be lost.

There is some evidence to suggest that a democratic form of organization leads to greater achievement on the part of all pupils and is to be preferred as a basis for educational advance (White and Lippitt 1960). The same arguments apply here as we used when discussing leadership of a team of teachers. If all are to be fully involved in the work of the group, all must share in decision-making. Furthermore, a democratic organization would seem to be the only form consonant with the social goals and principles that we have described as endemic to the mixed-ability class. Pupil groups, like teams of teachers, should be organized by democratic methods so that all members feel fully involved in their activities and can learn to take a share of responsibility for those activities.

Teachers should bear these factors in mind, therefore, and deal with situations where individuals tend to dominate the rest. To have a spokes-man for the group will often be convenient administratively, but to avoid the dangers of this, some kind of rotation of responsibility would seem to be necessary. Certainly it would seem desirable that a deliberate effort be made to encourage the reserved and modest pupils to take an active and responsible role from time to time.

The many different roles that the pupil is required and enabled to play in the class by this kind of approach to the organization of groups is well summed up in Elizabeth Hoyles' comments on the grouping of pupils at Vauxhall Manor School (1975, p. 61).

> One of the main features of mixed ability classroom organisation is the numerous occasions on which children need to work in groups. The child has to learn to use the group activity effectively from the beginning. First of all, she needs to learn how to play her own part in the group. Sometimes she will be the leader of the group, directing others, at other times she will be a member working at someone else's direction; sometimes she will be working with the group, sometimes she will be working away from the group. In addition, she will need to grow in judge-ment of other people's ability and this can join the general teaching the

school provides in respect for other people because, from an early stage, she will have to learn how to use other people's talents, not only for her own development but for the development of the group. It is important for her to learn how to apportion work to other members of the group so that when she becomes a leader she can use other people's talents to the full. In addition, there is a great need for learning tolerance since it will be impossible for every member of the group to fulfil exactly what the group expects and, therefore, she will need to learn to understand not only the abilities of others but also their failings and how they can best be helped.

Sociometric tests

If teachers are to be as sensitive to the subtleties of grouping as what has been said so far would seem to require that they should be, they will need a great deal of prior knowledge about all of their pupils. Teachers have been inclined to be more concerned with the intellectual capacities of their pupils and to pay less attention to the assessment of their social capacities, their sociometric status. Indeed, in a traditional class-teaching situation, such knowledge is not nearly so important, so that there has hitherto been little reason for them to be conscious of the need for it. Concern with the social education of pupils and with the problems of grouping them suitably in a mixed-ability class makes such knowledge essential and we must look briefly now at some of the techniques available to the teacher for obtaining it.

Following the lead of J. L. Moreno (1934), psychologists have shown that relationships within a group can be revealed by relatively simple methods. If we ask all the pupils in a class to tell us whom they would like to sit by or work with or spend a day out with, suggesting that they give us in each case two or three names in order of preference (it is not advisable, although it might be more revealing, to ask them to tell us whom they would not like to sit by, work with or spend a day out with), we can get a pretty clear picture of the social relationships existing in the class at that point of time. This information will only be given freely and truthfully if we assure them of total confidentiality and indicate that as a result they will have a chance to sit by or work with, if not spend a day out with, the person of their choice. From this information we can determine the sociometric status of each individual.

The information can be plotted on a simple chart, or sociogram, using a circle or some such mark for each pupil and arrowed lines linking the circles to indicate the direction of any attraction that exists in each case

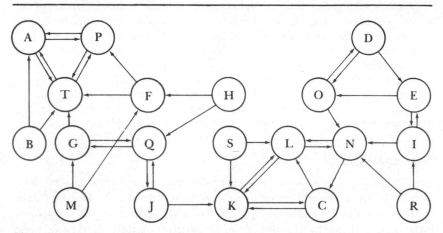

Figure 5.1   An example of a sociogram showing relationships within a class of twenty pupils

(see Figure 5.1). A clear picture then emerges of the sociometric status of each individual and the social relationships linking the individuals together. We begin to see who the isolates are, where mutual pairings exist, where there are triangular relationships and who the 'stars' are — those whose circle is surrounded by arrows pointing inwards.

The same information can also be tabulated by using a graph, or socio-matrix, if the teacher finds it easier to read in this form. Pupils are listed on both axes and relationships shown in the corresponding squares like, for example, a table indicating distances between major cities (see Figures 5.2 and 5.3).

Whatever form of presentation the teacher favours, however, this relatively simple test will have provided him with a great deal of the information he needs to help him with the supervision of the formation of groups.

Some words of warning must be put in here, however, before the reader rushes off to try this out. In the first place, this is only one of a number of ways in which one might set out to discover the social relationships existing in the class and it does tell us mainly about the social situation in relation to the particular question we have asked; the person a pupil

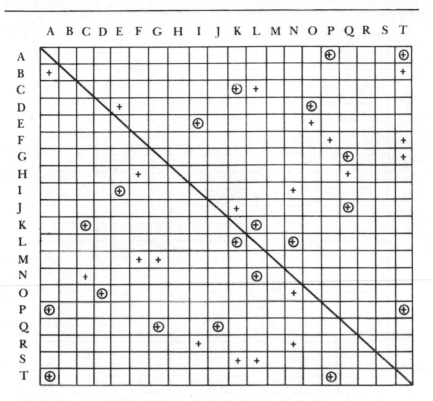

Figure 5.2   The same data expressed as a matrix. (+ = choice, ⊕ = mutual choice)

wants to sit by in class may not be the one he would like to work with or to spend a day out with.

Secondly, there is evidence to suggest (Northway 1968) that there is a high level of stability in the sociometric status of the individual, that pupils who are popular continue to be popular and that isolates continue to be isolated, but, although this kind of stability exists in the individual's ability to form relationships, the pattern of relationships actually formed is changing all the time as purposes change and as the work at hand varies.

In view of these factors, it is important for the teacher to remember that

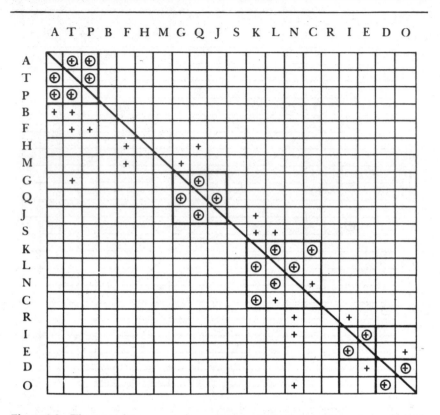

Figure 5.3 The same data expressed as a sociomatrix to show subgroupings within the class

what he has measured is the social situation of the class at the point of time at which he asked the question and in relation to the purpose embodied in the question — sitting by, working with, going out with or whatever — and that he needs to measure the relationships within the class regularly if he is to keep abreast with developments. Such regular retesting has the added advantage that it enables the teacher at the same time to measure the success of any attempts he has been making to improve the social status of his isolates and to keep an eye on the social development of each individual.

The changing nature of the relationships existing among any group of individuals is a further argument for that flexibility of grouping within the mixed-ability class which we have already advocated for a number of different reasons. It is also an argument for not being too rigid and scientific in our use of sociometric data. What we learn by using these techniques should guide but not control our decisions. It must be remembered too that the basic criterion we are recommending for grouping pupils is the area of their interests and, since academic progress continues to be a central concern, groups must be formed with reference to those as well as to social relationships. Here again the teacher has to play several instruments at once to achieve a harmony.

Summary and conclusions

In this chapter we have discussed the bases upon which the teacher might supervise the creation of subgroups within his mixed-ability class, emphasizing in particular the advantages of grouping pupils according to their interests. We noted again the need for that flexibility of grouping at this level that has been recommended elsewhere as a key criterion for the creation of the classes themselves within the school. Stress was also placed on the importance of recognizing the social purposes of our groupings, not least because our academic purposes will only be attained if we pay due regard to the social climate we create. We then considered some of the practical problems facing the teacher in creating the right kinds of grouping and finally it was suggested that he can gain some help from the use of simple sociometric tests to obtain one kind of data that he needs to create the most effective working groups.

Once these groups have been formed and the formation of them has been supervised along the lines that have been indicated, the teacher will need to get them going on their work by way of an organization something like that suggested in Chapters 3 and 4. He will need to spend time with all of them and this creates for him the practical problem of getting round to all in the time he has at his disposal, to give them the help that we suggested was needed. He can spend a brief time with all or he can devote more sustained attention to a few each time he has the class. The latter would seem to be the better policy but it will depend on the success with which he has been able to select and organize his resources, since the other groups will only be able to get on if they have at hand the materials and other resources they need. The organization of resources, then, is the teacher's next major problem.

CHAPTER 6

RESOURCES FOR TEACHING AND LEARNING

The story is told of how Periander, on becoming tyrant of the Greek city of Corinth in the sixth century BC, sent a messenger to the long-established and highly successful tyrant of Miletus, Thrasybulus, to ask his advice on the mode of government he should adopt. Thrasybulus said nothing, but took the messenger out of the city and into a field of corn and, as they walked through this field, he broke off and threw away every stalk of corn that rose above the rest.

As long as teaching has gone on, resources have been needed to aid and support it. That such resources have not always been successful is illustrated by the fact that when Periander asked his messenger on his return to Corinth what Thrasybulus had said, he answered that he had said nothing and Periander had to do a lot of questioning before any message came through. Not all teaching aids, however, are as deliberately obscure as this. Teachers have always been on the lookout for materials and apparatus that would help them to teach more effectively and their pupils to learn more quickly and more permanently. Most of the things they have used have been somewhat more manageable than fields of corn but there is nothing new about the use of resources for teaching and learning, other than perhaps the name.

However, interest in this problem of resources has in recent years increased considerably at all levels of educational discussion. Research in this field is being sponsored from several national sources including the Schools Council and the National Council of Educational Technology; many Schools Council projects have devoted a great deal of attention to the development of resources in connection with their own work; several local authorities, many individual schools, and a number of colleges and departments of education are building up resources centres, banks and collections; and teachers are becoming increasingly aware of the need to devote much of their attention to the development, use and storage of resource materials of all kinds.

There are two main reasons for this increase in interest. In the first place, technological developments have put a great deal more at the teacher's service and provided him with a greater variety of resource material to choose from. Traditional resources such as books, charts, pictures and so on are available to him in greater numbers and at a relatively cheaper cost than was once the case because of developments that have led to a greater facility of production. New and cheaper reprographic methods enable him to produce a wider range of materials himself. At the same time, the variety of resources available has been considerably extended by the development of new types of resource, again at a cost that makes them less of a luxury than was once the case. Such things as film-strips, audio- and video-tapes and transparencies for use with overhead projectors are now well within the reach of the average school budget, as is the 'hardware' required to make use of them, and teachers must now choose from such a large range of sources that making the right choice on educational and economic grounds has become a problem on a scale that could not have been envisaged in the days of chalk, talk and textbooks.

The second main reason for the recent increase of interest in resources is that curriculum changes and changes of organization have presented teachers with teaching situations for which the resources they have used in the past have often proved inadequate. A mixed-ability group, for example, cannot be adequately catered for by the provision of a copy for each pupil of all the basic textbooks; each will need to be provided with the books and other resources that will meet his own individual needs. This will be true even in a situation where it is the teacher's concern to deal only with one subject area and perhaps even one section of a detailed syllabus, so that there can be little doubt that the kind of approach to teaching we discussed in Chapters 3 and 4 will create even greater demands for resources of all kinds, demands that no teacher can ignore or easily satisfy. New kinds of teaching demand new resources and the particular approaches that have been introduced in recent years demand them both in greater quantity and in greater variety.

Uses and purposes

In order to obtain a clearer view of the kinds of resource that may be needed and a basis on which we can select among what is available, it may be helpful to begin by getting clear about the uses of resources and the purposes that they might help us to achieve. There would seem to be three main purposes that resources will be designed to serve. These three purposes we can look at separately but we must not lose sight of the fact that in practice they will be closely interrelated.

In the first place, resources have always been used by teachers as aids to motivation, as devices to stimulate the interest of the pupils. Few teachers can be unaware of the immediate transformation of the level of attention of any class, even in the most formal lesson, when an object is produced to illustrate a point that the teacher is making — a Roman nail found at Verulamium, a fossil, even a piece of coal or a lump of cheese. These aids are important in all lessons; they become vastly more important in a situation where we are trying to work from the interests of pupils. We have said elsewhere that the first task is to stimulate interest; this will be difficult to do successfully without a vast array of resources chosen for this purpose. Indeed, it was suggested earlier that one way of stimulating interest is to surround the pupils with resources of this kind and let them browse among them, explore them and find within them something that sparks them off. This would suggest that at this stage the resources may be more important than the teacher, whose first job is done when he has made a careful choice of resources and made them available to his pupils.

We must beware, of course, of overstimulating them so that they become merely excited and are not ready to settle down to the hard work that must follow. For once interest has been stimulated, as we have seen, the next task is to feed it and this is the second purpose of resources — to aid and promote learning once begun. Having gained some idea of the areas in which they want to work, pupils need to be able to start work as soon as possible. We suggested in Chapter 3 that the teacher should have work-cards ready to enable them to get started without his personal help, so that he can devote his attention to those who for one reason or another are not ready to begin work on their own. He will also need suitable resource materials that they can start work on.

Furthermore, having embarked on an enquiry, a project or a piece of work, pupils will need to be given access to sources of information and other materials that will enable them to work on it at an appropriate pace. We have already referred to the fact that interest will quickly be lost if it is not fed and in teaching through individual and group assignments it can only be fed by the provision of the right kinds of resource material. It is essential that this kind of material be available throughout the project so that the teacher does not need to cover the same ground several times with different groups and can save his time and energies for dealing with problems that must have his personal attention. It is important to remember too that material provided for pupils with this purpose in mind must also satisfy our first purpose, to maintain their interest and stimulate and give access to further interests. If a pupil has opened up a rich vein, we must provide him with the tools that will enable him to get all that is worth having from it and the incentive to keep working at it.

A third main purpose of resource material is to stimulate the imagination of pupils. In Chapter 3 reference was made to the dangers of always seeing the individual assignment in terms of information-gathering and of losing the opportunities that this kind of situation offers for engaging pupils in creative and imaginative activities.

It would not be appropriate here to become involved in discussions of what we might mean by 'creative activities' but attention must be drawn to the concern that has been expressed in recent years about the over-emphasis in schools on fact-finding, fact-storing and cognitive skills generally to the detriment of the development of creative abilities, and of the credit given to the convergent rather than the divergent thinker in the traditional structure of the education system (Getzels and Jackson 1962; Torrance 1967; Haddon and Lytton 1968). It has been felt not only that this results in an unbalanced education for some individuals, but also that it has serious implications for society which needs more creative people than the education system produces. The arguments here are strong. Clearly, for some teachers they will be of more direct and immediate relevance than for others, but few can ignore the demand that this places on them to be as aware of the development of the imagination of their pupils as of their intellect.

Nor must we make the mistake of assuming that these two aspects of the development of pupils can be treated in isolation from each other. We shall have cause to note again, when we come to look at the question of assessment in Chapter 9, that there is an inevitable interrelationship between feeling and learning, between the affective and the cognitive, so that when we set out to stimulate the imagination of pupils and arouse their feelings we are at the same time promoting their learning; conversely, we cannot seek to promote intellectual learning without giving any thought to the emotional factors involved. As we have just said, it is not possible in practice to distinguish the interrelated purposes of resource materials.

Work-cards

We have already referred several times to the fact that teachers of mixed-ability classes have found many advantages in the use of work-cards or work-sheets; they have become an increasingly familiar feature of many classrooms. Used in moderation and with proper care and thought, they have several advantages. They can and should be tailormade for the situation in which they are being used; they are an effective device for getting pupils quickly down to productive work without the need for the teacher to provide personal advice and instructions to individuals or groups; and, as we shall see in Chapter 7, they offer scope for varying the level of work to suit different abilities.

Before considering some of the problems associated with the production or collection of other types of resource material, then, it may be worthwhile spending a little time discussing the use of work-cards.

One way of using them to some effect and many of the principles to be borne in mind when preparing them emerge clearly from the account given by Phil Prettyman of the use of work-cards in the teaching of mathematics to mixed-ability classes in the first year at Crown Woods School (1975, pp. 154-160).

> After much discussion it was decided to arrange first year content into fifteen or so topics, each containing one or two basic concepts, and to write a set of work-cards on each topic. Topics chosen covered a wide range such as Plotting Points, Symmetry, Statistics, Fractions, Angles, Mappings and Relationships, but were largely chosen so as to present a good selection of new ideas and concepts to children without arousing any fears and inhibitions from their Primary schools.
>
> Each topic is presented as a set of work cards. The number of cards in each set varies considerably but the usual number is eight or nine. In all cases the aim of a set of cards is to present the conceptual idea or principle in as simple a form as possible within the first two or three cards. On these

cards the language and style are as simple and straightforward as the topic allows. The succeeding cards then practise and develop the idea or technique at a level with which the majority of children can cope. The set is then completed by a couple of cards in which the topic is given a bit of depth with the style and language correspondingly more complex.

Clearly the effectiveness of this approach depends very much on the way this material is actually used, and . . . there is a range of different ways in which this work can be used in the classroom.

This then is the pattern on which we have tried to model all our first year work. To illustrate this, here is an example of one of the topics. Entitled 'POINTS' it has the objective of teaching children that any point on a plane can be determined by an ordered pair — Cartesian co-ordinates.

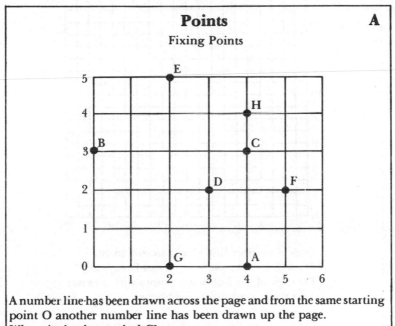

A number line has been drawn across the page and from the same starting point O another number line has been drawn up the page.
Where is the dot marked C?
To get to C from the starting point go *along* 4 and then *up* 3.
C is the point (4,3)
Notice that the *first* number in the bracket is the *along* number and the *second* number is the *up* number.
(4,3) is called an *ordered pair*.

Write down in your book the pairs of numbers which fix the positions of A,B,C,D,E,F,G,H.

Card A establishes the concept. It explains as simply as we have been able to make it that a point on a grid is fixed by a pair of numbers.

A few children may need some verbal support, but only those whose reading is weak, and the rest of the class grasp the idea very easily. Some teachers prefer to replace card A by a classroom discussion which can have the effect of stimulating the interest of the children more effectively.

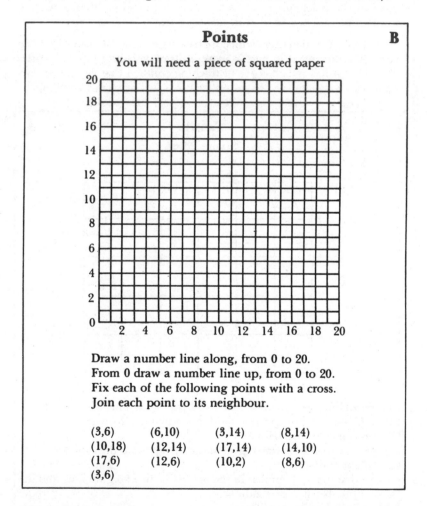

Points   **B**

You will need a piece of squared paper

Draw a number line along, from 0 to 20.
From 0 draw a number line up, from 0 to 20.
Fix each of the following points with a cross.
Join each point to its neighbour.

(3,6)	(6,10)	(3,14)	(8,14)
(10,18)	(12,14)	(17,14)	(14,10)
(17,6)	(12,6)	(10,2)	(8,6)
(3,6)			

This is the first of several cards which get the children to draw a shape by joining dots up in an order. A simple symmetrical shape like this is good for a start since it assists the children to plot the points correctly.

A later card which exploits the same idea is card H.

Points   H

Use Squared Paper

Draw a number line from 0 to 28 along.
Draw a number line from 0 to 28 up.
Plot the following groups of points.
In each group, join each point to its neighbour.

Group 1

(8,13)	(7,12)	(8,11)	(9,11)	(12,16)
(11,18)	(7,17)	(6,17)		

Group 2

(18,12)	(18,9)	(20,6)	(23,4)	(23,2)
(22,1)	(25,1)	(25,5)	(24,6)	(25,7)
(26,10)	(27,7)	(27,11)	(25,16)	(18,19)
(9,18)	(6,19)	(3,17)	(2,15)	(2,3)
(3,2)	(4,2)	(5,3)	(5,4)	(4,4)
(4,3)	(3,3)	(3,4)	(4,9)	(6,10)
(6,9)	(7,10)	(8,8)	(9,4)	(9,2)
(8,1)	(11,1)	(11,7)	(12,8)	(12,9)
(11,10)				

Group 3

(12,8)	(13,7)	(17,6)	(20,6)

Group 4

(5,14)	(5,15)	(4,15)	(4,14)	(5,14)

This card is highly successful, and although there ar four cards of this type in the set, the children very much enjoy using them.

The cards between B and H use battleships, a map, letters of the alphabet, all to reinforce the same idea.

Card J extends the idea of plotting points into plotting straight lines and suggests that there might be a relationship between the coordinates.

Points   J

Draw number lines from 0 to 20

1
Plot these points. Join each one to its neighbour:

(1,2) (3,6) (5,10) (7,14) (10,20)

What shape do you get?

Can you see a connection between the along number and the up number for each of the points?

2
Plot the following sets of points. Draw each set on a different piece of squared paper. In each set, join each point to its neighbour.

(a)	(1,1)	(7,7)	(9,9)	(10,10)	(15,15)
(b)	(3,1)	(6,2)	(12,4)	(15,5)	(18,6)
(c)	(1,8)	(4,11)	(6,13)	(8,15)	(10,17)
(d)	(8,3)	(9,4)	(12,7)	(16,11)	(20,15)

Can you find a connection between the along number and the up number for each set?

K again explores another idea that would normally be met at a much higher level.

Points   K

Plot on squared paper the following pairs of points and join up each pair with a straight line.

(a)   (2,4)   and   (2,6)
(b)   (5,3)   and   (1,3)
(c)   (2,9)   and   (6,5)
(d)   (8,0)   and   (0,2)
(e)   (1,2)   and   (5,8)

Write down the positions of the middle point of each of the lines. Example — the middle point of the line (a) is (2,5).

Try to find out a way of knowing the middle point of the line joining (8,3) and (6,5) *without* plotting the points.

Look carefully at the answers you found for (a) to (e).

Clearly, not all the children may complete the whole set of cards but this will not matter so long as the faster children reach the end when the slowest has completed the introduction and is on the practice — the principle has been learned.

As with all our work this set of workcards was hand written out on a spirit master and enough copies for three or four sets run off. At the end of the initial year this, and all the other topics, was re-discussed and edited to produce a better finished final version. This version is written with Letraset and indian ink then electro-scanned to produce a master stencil: this technique, if the master copies are filed, can produce an almost infinite supply of copies (paper supplies permitting!).

For each topic the teacher receives an envelope file containing a stack of transparent plastic envelopes, each containing thirty-five copies of one worksheet. Each child also has a transparent plastic envelope in which he places his current worksheet. This system protects the worksheets and gives them a reasonable life-span.

Some teachers have discovered the advantages of also producing work-packs to be used in conjunction with work-cards. A work-pack will provide a pupil or a group of pupils not only with a list of tasks to perform or questions to seek answers to but also with some of the sources of information, materials and even equipment needed to fulfil the work-card's requirements. A useful device in the infant school, for example, for getting children involved in elementary scientific exploration is to provide a pack which will start them off on an exploration of 'What things float?' — a bowl for water and a number of different objects and kinds of material — or one that will encourage them to discover 'What things are magnetic?' — a magnet and pieces of different metals and other materials. Even the teacher of average ingenuity can quickly think of simple collections of items that can be made in this way not only to provide the pupil with a range of tasks but also to help him get started on the work that will enable him to accomplish them. Some packs developed in connection with particular curriculum projects and some commercial products, such as those packs which contain copies of historical documents and other sorts of evidence, offer similar advantages, but it is not necessary to rely solely on prepackaged products.

Many subjects or topics will, of course, require the further support of other kinds of resource material and we must now turn to a consideration of some of the problems associated with the collection, use and storage of these.

Choosing resource material

To help teachers with all aspects of their teaching there is nowadays a wealth of material available, some of it prepared directly for them by publishers and others, some of it available from the things we all come across in our everyday lives. Some of it is in straightforward graphic form — cuttings from newspapers, comics and colour supplements, posters and so on; some will be materials that can be looked at, handled and perhaps manipulated in various ways — rock samples, geometrical shapes and solids, models designed to demonstrate certain scientific principles; others again will be of a kind that will require special equipment for their use — audio- and video-tapes, films, film-strips and slides. Nor must we lose sight of the fact that places pupils might visit and people who might visit the school or be visited by groups of pupils can be important sources of learning.

There is nothing that *prima facie* is not potentially a source of learning or an aid to teaching, so that teachers can become great collectors of junk. The profession has its Steptoes, as we all know, but some criteria of selection are needed if teachers are not to become buried under a mass of unused materials and certain conditions seem to follow from the basic purposes we have outlined.

A prime consideration is that the material offered to pupils should be the most interesting that is available. To some extent, decision on this is a matter for personal preference and judgement and, in any case, it will often be possible for children to select for themselves and teachers will have this very important feed-back to help them with their choices. One general principle, however, does seem important and can be offered as a broad guide to teachers choosing material. Preference should always be given to first-hand resources where they are available. A piece of bark will always be a better aid to learning than a picture of a piece of bark; a newspaper cutting on a particular topic will always be better than a statement about it prepared for use in the classroom; a visit from the local 'bobby' will always elicit more interest and learning than exhibiting his helmet, badges or truncheon. To say this is, of course, merely to draw attention to a general educational principle that applies in all teaching situations, but its importance in the selection of resource materials must not be forgotten.

A second vital point is that the resources we prepare for pupils should satisfy the educational criteria we discussed in the first chapters of this book. They should be related to the goals and principles of the project they are meant to serve; they are not to be seen merely as devices for keeping pupils busy, but as aids to the achievement of whatever purposes the teacher or pupil has or comes to have. They must ensure that the pupil is stretched. We have stressed the need to avoid undirected learning or a 'hobbies' approach and the resources we provide must make appropriate demands of each pupil. They should also be chosen with the development of understanding in mind. We have claimed that this, rather than the acquisition of knowledge, is the proper aim of education and the resources we choose should reflect this. Furthermore, if we believe that the road to education may be different for each individual pupil and that we should individualize our educational provision, they should reflect this too. Finally, we should not lose sight of the need to promote the kind of imaginative and creative work we have just been discussing, to attend to the affective development of our pupils by every means available to us.

This leads on naturally to a further point that must be made about the criteria we should use in selecting resource material. The quality of that material should be such as to promote whatever standards of accuracy, expression, presentation and so on we wish to see reflected in the work of our pupils. If we offer them tatty resources, we will have no grounds for refusing to accept shoddy work from them in return. This is a particularly important point to bear in mind when preparing worksheets. If teachers are to produce these themselves — and there are obvious advantages if resources can be tailormade in this way to suit the particular context of an individual school or classroom — some attempt must be made to achieve a reasonable level of professionalism in their production. Even if accuracy of typing cannot always be guaranteed, errors of spelling and punctuation cannot be condoned.

A third important consideration is that resources to be used by the pupils themselves must be of a kind that they can use, for the most part, without help from the teacher. Obviously, this means first that they must be suited to the pupils' level of understanding in terms of the language in which they are expressed or the illustrations, diagrams and the like that they may contain. This in turn implies a great variety of materials and also a wide range of books, since not all pupils in a mixed-ability class will be able to profit from the same material. As we shall stress when we come to consider the slow learner and the nonreader in Chapter 7, it means that it will be necessary to prepare our work-cards and many of our resources in a number of forms and at a number of different levels of complexity. Nonreaders, for example, or those whose skill in reading is very limited will need to have reources and work-cards prepared for them in different media.

Clearly, the audio-tape can have particular value here, but it may not be so easy for the pupil to use and this brings us to another major consideration. Resources must be in such a form that they are readily accessible to the pupils. In many cases this may mean that they will have to be restricted to items that can be used directly without the aid of any special equipment, but teachers are increasingly coming to feel that since such equipment is now more readily available than it once was, its use should no longer be restricted to the teacher, who can only use it in a mass showing to the whole class, but it should be made available for individual pupils or groups to use, as and when it seems appropriate for their work. The language laboratory is now an established part of the educational scene in many secondary schools. There is no evident reason why the advantages that language teachers have been able to gain from the use of new equipment by pupils should be denied to other teachers. A classroom devoted to individual or group learning can readily include areas for viewing video-tapes, film-strips, films or slides and for listening to audio-tapes, although in a team-teaching project it may be better to have each kind of machine in a different room. Furthermore pupils who are accustomed to operating television sets, radios and record-players in their homes and often tape-recorders, movie-cameras and projectors of various kinds too, can be trusted to operate similar machines in school — perhaps more than many teachers.

Some pupils, then, will know how to use equipment of this kind; the rest will need to be shown. This can be done quite easily by a teacher who knows his way around the basic educational hardware. This in turn highlights the fact that an introduction to the workings of educational technology is nowadays an essential part of any teacher's training. Whatever we feel about the desirability of allowing pupils to operate these machines themselves, there is no doubt that teachers must have this kind of expertise.

It must also be stressed, however, that teachers need rather more than a course on how to work the machines; they also need a more extensive training in the uses and applications of educational technology and in the preparation of their own materials for use with the machines, since here again the first-hand item, prepared for a specific task, is infinitely preferable to the 'shop-bought' model.

Teachers trained in recent years should have been given this kind of training; older teachers will have had little preparation for work of this kind and their experiences of educational technology may have engendered in them a great suspicion of all such aids. They will need to be persuaded that there have been great advances in this field since the days when Olympic high-divers leapt gracefully feet-first from swimming bath to high-diving board and Mexican cowboys galloped at great speed rump-first across the pampas of South America to the accompaniment of a growling soundtrack which eventually brought everything to a complete standstill. The level of sophistication in educational technology is much higher now, as, to be fair, many older teachers seem to have discovered, since there are indications of a greater use of educational technology by older teachers than by the younger ones.

In time, then, it must be hoped that the greater variety of resources now available to teachers can be made available also to pupils in this way. But it is also to be hoped that neither teachers nor pupils will be misled by the nature of these new developments into falling for gimmicks. The emphasis must remain on the educational aspects of educational technology and the technology must not be embraced for its own sake. Its advantages can only be evaluated in terms of its contribution to the advancement of the learning of the pupil and in relation to the contributions to be made by other kinds of resource.

In particular, we must not forget the advantages of books, one of the oldest but still one of the most reliable and accessible of resources. In this connection it is interesting to note that where other resources have been made available to pupils an increased interest in books has seemed to result. An extensive range of books of all kinds should be made available, perhaps through some kind of class library scheme, and no opportunity should be lost to ensure maximum use of the resources offered by the school library. In fact, planning from the start has to be done in close cooperation with the school library, since demands on it are necessarily increased enormously once individual and group assignments are begun.

A resources collection

Whatever kinds of resource material the individual teacher favours and is able to use and make available to his pupils, there is no doubt that work with mixed-ability classes, and in particular work that is enquiry-based or interest-based, will make it essential that he amasses great quantities of material. A single project will require a great mass of material of all kinds; a team of teachers will collect even more. Much of this material will be of a kind that can be used over and over again with successive groups of pupils and, although it is obviously desirable that teachers should be continually adding new items and weeding out the outdated to ensure that it is always abreast of the times, it is clearly not good sense to start from scratch each time and develop entirely new resources for each project. Some method of storing material that has been collected is vital.

Most teachers will, of course, keep material that they feel has a value for the future in a cupboard and the whereabouts of each individual item in their heads, but in time one of two things will happen. Either they will go to the cupboard and fail to lay hands on the item they want, perhaps in a situation where a pupil or group of pupils is crying out for it, or they will one day open the cupboard door and be buried in a deluge of colour supplements, film-strips, slides, specimens, pictures of Hannibal crossing the Alps and other such bric-a-brac — an experience unpleasant to the teacher, although highly entertaining for the pupils if they happen to be around when it happens. It is at this point that the teacher realizes that he needs a resources bank or collection. For what converts a cupboard full of junk into a resources collection is the introduction of a system of indexing and storing which will ensure easy access to, and retrieval of, each individual resource item collected.

In the first instance, such a system need not be highly sophisticated (Hanson 1971). The resources collection at my own college began with a few box files in a cupboard and a shoe box for the index cards (Ellis 1971). It is enough to get hold of several boxes of a similar size, either purchased for the purpose or begged from a local grocer, so that items can be separated under broad headings. Some sort of simple card-index system will enable one to extend this and will help with the retrieval of items stored. A card for each item, on which is indicated, first, the broad category it has been allocated to and, therefore, the box it is kept in and, secondly, a number to show its whereabouts in the box, is simple to prepare and, if kept in alphabetical order with other such cards, affords quick and ready access to any item. A further refinement of some value can be achieved by varying the colour of the index cards — for example, white for cuttings from newspapers and magazines, yellow for slides, pink for posters and charts and so on.

The introduction of a system such as this will bring immediate improvement and a feeling, for the most part justified, of being in control of things again, but it does have serious limitations. Items will still be 'lost' or much time wasted in the retrieval process because of the difficulties of classification and the need for some form of cross-referencing. For example, a cutting from a colour supplement describing the plight of women and children in the Indo-Pakistan war might be classified under 'Women', 'Children', 'Family', 'Hunger', 'War', 'India' or 'Pakistan'. Whichever heading it is listed under, it will not be turned up by anyone looking for items on any of the other six topics, although it might be highly relevant to them. Furthermore, anyone looking for items concerned specifically with, say, 'Women in War', will waste time looking at a lot of items under both 'Women' and 'War' when in fact he is interested only in those that link both of these classifications together.

Some system of cross-referencing, therefore, becomes important at a very early stage in the development of a resources collection. One answer to this problem, and one favoured by many, is to keep several cards on each item and thus list it under all the appropriate headings, each card containing also an indication of the other headings under which the item is listed.

A more sophisticated system is the optical-coincidence punch-card system. With this system a punch-card is prepared as an index card for each topic and item number 672, for example, which is our cutting about women and children in the Indo-Pakistan war, is recorded by a hole punched at square number 672 on all the relevant index cards — 'Women', 'Children', 'Family', 'Hunger', 'War', India' and 'Pakistan'. That 672 is an item relevant to the theme 'Women in War' or 'Women in War in India' will be quickly discovered when we take the index cards, 'Women' and 'War', and 'India' too if we wish, and place them on top of one another. The holes at number 672 will clearly coincide, indicating the relevance of this item, whereas items relevant to only one of these themes will not show up. Thus quick retrieval of only those items that are fully relevant to our particular purposes becomes possible. Equipment can be obtained for easy reading of these cards, but they can be read without such elaborate equipment.

Such a system is, of course, at a level of sophistication some way beyond the shoe box we started with. It is also seen by some to have drawbacks which outweigh its advantages. It is more complicated to run than the simple card-index system and it makes it very difficult to withdraw items from the collection when they become outdated, since although the holes in the punch-cards can be filled in or some other method of showing withdrawal used, this is not so easy as simply destroying the relevant card.

Whichever system is favoured, the crux of the problem is the categories that are used for classification. To a large extent, material should be classified under headings that are relevant to the courses or projects they will be required to serve. Since we are working from the assumption that the problem arises from the fact that masses of material have been collected already to deal with earlier projects, we can begin by saying that it should be classified under the headings under which it was first collected, since these clearly were the relevant categories then. A list of themes should be drawn up primarily on this basis, but such a list should always be open-ended so that other categories of classification can be added as and when they seem appropriate and necessary.

For a number of reasons, however, teachers would be well advised not to make such a system too idiosyncratic. So far, we have treated this problem as one that is of growing concern to the individual teacher facing up to the requirements of his mixed-ability class either within his own specialist subject or across a range of subjects he might be responsible for. As each teacher's collection of resources grows, however, it will prove uneconomical in every sense for every teacher to continue to develope his own collection and some kind of centralization will become essential. Furthermore, we have already seen how the new approaches to teaching that a mixed-ability form of organization makes necessary often lead to some form of team-teaching and this again makes it necessary to think of developing a joint resources centre which all members of the team can use. This kind of development should be borne in mind from the outset if a smooth transition is to be effected to a large and more central organization of resources, so that some generally agreed basis of classification is needed at the beginning.

Ultimately, such a resources centre must be closely linked with the school or college library and this development must also be kept in mind from the very beginning. Even in the early stages of a resources collection, there are enormous advantages in cataloguing not only items actually contained in the collection but also other relevant items to be found elsewhere. The inclusion of references to basic books in the school library is, therefore, a great asset. In this way, a link with the school library can and should be maintained throughout, so that there would seem to be certain advantages in adopting a system that is as compatible with the Dewey system of classification as is possible. This system of classification itself is not always the most helpful, since it was not designed for teaching purposes, and teachers must not allow themselves to be kept too strictly to it, but to keep as close to it as possible will help when the resources centre becomes a part of the school library.

When this happens, we will have progressed a long way from our shoe box and cupboard beginnings. For what we can now hope to have is a Resources Centre (note the transition to capital letters) in its own purpose-built accommodation. Such accommodation should include, in addition to storage space and places for looking at pictorial, written and other nonprojected materials, booths for the viewing of projected materials of all kinds — slides, film-strips, films, records, audio- and video-tapes. It should also include facilities for making and reproducing items, since teachers need to be able to produce their own teaching and learning aids, or have them produced for them, and pupils need to be able to secure copies of items they may need in their work but of which the Centre may contain only reference copies — there is no point in loaning a precious copy of an article if this can be reprographed and the original retained in safe keeping. Facilities for photocopying, typing, duplicating must be readily available, therefore, and there must also be provision for the recording of radio and television broadcasts and the preparation of other audio- and video-tapes, of films, slides, transparancies for use with overhead projectors and any other aids to teaching and learning that teachers or pupils may need.

Such provision may seem a long way off in most schools, but it is some-thing to be worked for and a clear vision of what the future may bring will help to ensure that the early stages of the development of a resources collection are approached in a way that makes subsequent development along these lines as smooth as possible.

Most schools are still at an early stage in the development of their collections of resources, but since this is an area of growing need, many people have already moved into this field and offer ready-made collections or packs of resource materials on particular themes or areas of work. Several Schools Council projects have seen the development of resources as an essential element in their work and have prepared resource packs for sale to schools. Others too have been quick to see the need and to prepare collections of material for use by teachers in certain areas of work. Indeed, some publishers are seeing the need to divert some of their attention from the preparation of textbooks to the publication of packs of materials of this kind. All of these sources will provide the teacher with assistance in his basic task of feeding the educa-tional appetite of his pupils and it must be acknowledged that for the most part such large-scale undertakings have the advantage of being based on wider research and experience than is possible for the individual teacher.

However, it must again be stressed that each school should develop its own individual collection suited to its own unique needs. What is generally available may be based on sound research and a resultant knowledge of general needs and may have much to offer the individual teacher, but the individual teacher has a knowledge of the individual pupils he is catering for as well as of his own teaching style and to rely entirely on material prepared for him by others is to lose the advantage that such personal knowledge affords. This is one aspect of the current recognition of the value of school-based curriculum development to which we referred in Chapter 1 as providing 'more scope for the continuous adaptation of curriculum to individual pupil needs than do other forms of curriculum development' (Skilbeck 1976, pp. 93-94). If, as we are continually suggesting, a major feature of mixed-ability grouping is an individualizing of education, then clearly our curriculum development must be school-based and, as a practical result of this, our supporting resources must be personalized.

Furthermore, there are additional advantages to be gained from a situation in which the users of the collection are also contributors to it. For in addition to the fact that teacher-users can thus develop a highly personalized collection of material to assist their own teaching and the development of their work, there is no better guide to what is good resource material nor any better criterion of selection than the observation of what pupil-users do in fact use. Fundamentally, a resources collection should begin from the need to store what has been used by the pupils. There is nothing that is more 'first hand' than that and teachers should not lose sight of this fundamental point in their desire to provide more and more sophsticated material. Nor should pupils always be 'spoon fed'. There is much to be learnt by being required to do one's own research in the full sense of the word, to find one's own source material instead of merely using other people's, and there is no greater assurance of the value of one's work than to see material one has obtained entering the resources collection for future generations to use.

There can also be profit in encouraging pupils to assist with the actual classification and indexing of materials. It is always a temptation, of course, to take advantage of the cheap labour pupils can provide, but it is possible to justify employing them in this way on educational grounds. In a college of education this work should certainly be done by students, since this is another skill they should learn in their preparation for teaching, but this is not the only advantage. The problems of collecting, sorting and classifying can be seen to have a certain educational value in themselves, since a good deal of careful and analytical thinking must go into them. To adapt the saying of Rousseau, 'While he thinks himself a librarian, he is becoming a philosopher.'

Summary and conclusions

In this chapter we have discussed some of the varied kinds of resource material currently available to the teacher and the main criteria by which he should make a selection among them. In particular, stress has been placed on the need to be clear always about what we are hoping resource materials will help us to achieve and to avoid the use of new devices and aids simply for their own sake and without reference to the precise advantages to be gained from them. In this connection emphasis has been laid throughout on the special merits of those resources the teachers collect or make themselves, since only these will be tailormade to fit the particular context of their own teaching, a point that was illustrated by the example of one set of work-cards prepared for a particular school situation. Finally, consideration was given to some of the problems of the storage of resource materials and the building up of a resources collection, and attention drawn to the advantages of establishing such a collection from the outset on a basis that would make it easy for it later to grow into or be absorbed by a full-scale Resources Centre.

The provision of resources for learning, then, is another major task for the teacher of a mixed-ability class and perhaps more complex when considered in detail than it appears to be on a first and rather superficial view. The important thing is to be clear about the general principles upon which one's own collection will be based and on the techniques that should be used to ensure that no opportunities for future development on whatever scale are lost. If the individual teacher is clear about these, then he can develop a system that can be adapted to any level of sophistication of materials, equipment or indexing and which will help him with the task of providing the right kind of education for each of his pupils.

CHAPTER 7

PUPILS WITH LEARNING DIFFICULTIES

One of the most interesting features of the early history of streaming, as we saw in Chapter 1, is the extent to which its introduction was prompted as much by a concern for the well-being and progress of the 'less able' pupils as by any consideration of the needs of the more gifted. Mention was made in Chapter 1 of the impact of the work of psychologists on the organization of the education system at that time and we noted in particular the influence of the work that was being done on the nature and causes of backwardness by men such as Cyril Burt. As a result of this work, it was felt that the needs of such pupils could best be met if they were kept apart from their brighter peers and given special provision in special classes. Such an arrangement was prompted mainly by a concern to cater for their academic needs in what seemed to be the most effective manner, but it was also assumed that it would not be good for their emotional development if they gained too clear a sight of the intellectual superiority of many of their peers.

This view persisted for a long time. In Daniels' study of teachers' attitudes to streaming in junior schools in the late 1950s, to which reference was also made in Chapter 1, almost half of the teachers who were asked stated that they would use streaming as a device for creating a smaller than average 'C'-stream class (Daniels 1961a) — a clear indication that they saw it as a system that offered advantages of this kind in the education of children with learning difficulties. Similarly, Brian Jackson's study in the early 1960s revealed that of those teachers who favoured streaming 95% did so because, among other reasons, 'a prime purpose of streaming is to help the less gifted child' (Jackson 1964, p. 32); in fact this reason was given by more teachers than any other. There are still many teachers, especially among those concerned with 'remedial education' in secondary schools, who favour streaming and thus oppose mixed-ability grouping on these grounds.

There may be some force to such an argument when we are considering the needs of those pupils who are in fact allocated to 'remedial classes'. Certainly, such pupils are provided with the kind of very special attention they are deemed to need, usually by teachers who are specially trained and often highly gifted in the sensitive task of encouraging and helping such pupils.

However, there are at least two factors that must be weighed very carefully before we acknowledge that the needs of these pupils make a streamed form of grouping desirable. In the first place, we must remember the plight of those pupils who are in the middle of the 'ability range', or towards the lower end of the middle, those pupils who may have certain kinds of learning difficulty but who do not quite come into the 'remedial' category — the 'B' stream or the lower part of the middle 'band'. These pupils do not usually receive specialist help of any kind and thus gain no special advantage from working in a streamed class with pupils supposedly of the same level of ability.

Secondly, we must take full cognisance of the growing body of evidence concerning the relative effects of streaming and unstreaming to which we referred at some length in Chapter 1. There is no evidence that the level of attainment of the 'less able' pupil is in any way higher in a streamed 'remedial' class. Indeed, such evidence as there is would suggest the contrary (Daniels 1961b; Barker-Lunn 1970; Newbold 1977). On the other hand, as we also saw in Chapter 1, there is evidence that clearly points both to the detrimental effects of streaming on the emotional and social development of the 'less able' pupil (Hargreaves 1967, 1972; Lacey 1970) and to the positive advantages in this area of un-streaming (Barker-Lunn 1970; Ferri 1972; Newbold 1977). It is apparent, therefore, that to separate such pupils from their fellows is to exacerbate rather than to alleviate their difficulties and those of their teachers and this consideration has often been a major factor in decisions to change to a mixed-ability form of organization.

However, there is little doubt that this is a feature of the mixed-ability class that worries teachers more than any other when they approach such a class for the first time. This is particularly the case for those teachers in erstwhile grammar schools who, as a result of the comprehensivization of their schools, are facing pupils of average and below-average abilities for the first time in any kind of professional context. We must carefully consider, therefore, how the teacher can best resolve the difficulties and take advantage of the opportunities the mixed-ability class offers him in this sphere.

Some aspects of this problem

The mixed-ability class gives the teacher the chance to let all pupils work at the pace most suited to them and at the level most suited to them. The pupil who finds learning difficult does, however, provide problems in this situation. He will probably lack the confidence and some of the very basic skills that might be felt essential to individual or group assignment work at secondary level, he may well have acquired, by the time he reaches the secondary school, certain emotional and social problems which the teacher will need to attend to and he may, as a result of persistent failure, if he has been made to feel conscious of it, have become a problem pupil, providing the teacher with problems of behaviour and control. Such a pupil can be troublesome even in his first year at a secondary school and, unless a change is made in the nature and quality of his educational experience, he will, as any teacher knows, give particular trouble in the last year of his schooling. While it is also true that the gifted pupil can be difficult if bored, to a large extent he can be dealt with by overcoming his boredom; it is the pupils who have learning difficulties that the teacher needs most help with in the mixed-ability class.

Some general points need to be made about such pupils before we can begin to consider how we might cater for them in a mixed-ability class. To begin with, it is important to be clear about the extent of the problem. It is difficult to be precise about this, not least because of the problems that surround the establishment of adequate criteria of measurement in such a problematic area. There is general agreement, however, that whatever criteria one employs, about 10 percent of pupils will be found to be significantly backward in the basic skills of language and number (DES 1964). Furthermore, although the figure will vary from school to school and from neighbourhood to neighbourhood, there can be few nonselective secondary schools which do not contain a significant number of pupils who are nonreaders or very poor readers. The problem, then, is widespread and is one that every teacher of a mixed-ability class will experience.

The most important point to be kept in mind when looking at this problem is the changes that have occurred in recent years in the view taken by psychologists of the nature of intelligence. The old concept of intelligence as a fixed and unchangeable level of general ability with which the individual is born and which is determined, therefore, almost entirely by nature has given way, as we saw in Chapter 1, to the view that, although nature may place certain end limits on its potential development, nevertheless intelligence can be developed by education, that we have no way of knowing in advance or predicting what the end limits of such development are, that childhood experiences can promote or retard its development and that nurture is at least as important here as nature. I suggested in Chapter 1 that this was what Lord Boyle had in mind when, as Minister of Education, he prefaced the Newsom Report with the now famous dictum that 'all children should have an equal opportunity of acquiring intelligence'.

This is one of the basic reasons for the change to mixed-ability classes. It is felt wrong to label children at an early age, to 'discover their IQ' and offer them an education appropriate to it, since there is evidence to suggest that nothing determines a child's IQ so effectively as the teacher's own view of what it is and the expectations that result from the teacher's view of it (Pidgeon 1970; Downey 1977). This is the 'self-fulfilling prophecy' to which we have already referred. If we only demand a 'C'-stream level of work from pupils, this is all we will get and we will not then be giving them the opportunity to 'acquire intelligence'. If intelligence can be developed, there is scope for the teacher.

There is similar encouragement for the teacher in the views of those psychologists such as Piaget and Bruner who have for the most part abandoned this rather passive view of intelligence and have concerned themselves with a study of intellectual development, adopting, as we also saw in Chapter 1, a developmental rather than a psychometric concept of intelligence and directing their attention towards how children think and how the quality of their thinking can be improved rather than attempting to measure intellectual ability as if it were a fixed quantity. This approach has led to a view of education as being concerned to accelerate intellectual growth rather than to purvey information and has taken it as a basic assumption that intellectual development can be promoted by the right kind of educational provision. Again there is scope for the teacher. The mixed-ability class is one way of helping him to use this scope.

Another reason why streaming has come to be regarded as an inefficient method of grouping pupils with learning difficulties is that it assumes that it is possible to generalize about such learning problems and thus it inhibits the making of more subtle distinctions between different types of learning difficulty. A child may be a slow learner in only one or two areas, but, if placed in the 'C' stream, he is offered a 'C'-stream diet in all areas and is thus undernourished in some.

A distinction needs to be made, therefore, between the kind of learning difficulty that is the result of what appears to be a limited intellectual potential and that which results from a failure to make effective use of an ability that seems to be present. In other words, one must be aware of the different types of provision needed for those pupils who are 'dull' and appear to reveal a limited ability or potential in all areas and for those who are 'retarded', who seem to be underachieving or under-functioning in particular areas, when their work in those areas is compared with what is known of their general potential from the level of their achievement in other fields.

The latter require help that can be truly described as 'remedial', since a particular and, one hopes, temporary deficiency is to be made good and the provision of such help can be seen to be possibly a short-term measure to which they may respond quite quickly. The former need to be viewed in a long-term way, since the problem here is not to put right some temporary deficiency, but to provide them with a complete education, whose broad goals must be the same as for all pupils, but which must be developed in a way that makes best use of their strengths and must be designed to make the most of their apparently limited potential.

However, although each of these groups needs a different kind of provision, we should never be in too much of a hurry to place individuals in one group or the other, since once we have classified a particular pupil as 'dull', we will be inclined to restrict the demands we make of him and we should always be reluctant to do this, since, as we have seen, teacher expectation is a key factor in the achievement of all pupils. We need to look carefully into each case to discover the reasons for particular learning difficulties and we need, therefore, to have some idea of the kinds of factor that seem to cause such difficulties.

Clearly, limitation of intellectual potential is one major cause of learning difficulties, but there are other reasons which are often more crucial. We have already referred in discussing the grouping of pupils to the importance of the emotional and social environment in which pupils learn for their intellectual progress. There is clear evidence that learning difficulties are often due to emotional and social deprivation (Tansley and Gulliford 1960). The absence of a mother's love in the early years of life may be as serious a disadvantage to the individual's subsequent intellectual development as we know it is for his emotional development and the continued absence of a secure and stable background in the home will reinforce this initial disadvantage.

The educational level of the home too will affect a child's rate of progress at school. The education of the parents themselves and their attitudes to education, the kind of language used in the home, the presence or absence of books, papers, records and so on will have serious implications for the child's 'readiness' for education and his ability to profit from what the school offers him, since factors of this kind will govern his level of motivation and the skills and attitudes he brings to school with him.

The pupil's physical capacities must also be taken into account, since they too have a direct bearing on his educational progress. At one level we must be on the look-out for physical disabilities such as partial deafness or sight difficulties that can be root causes of learning difficulties if not detected in time, but at another level, we should also be aware of the implications of physical maturity for the intellectual development of all pupils. A certain neurological maturation must take place before the individual can learn certain things and there is a clear link between the kind of intellectual development analysed by Piaget and others and the physical development of the individual. This seems to be particularly significant at the adolescent stage where the pupil who is late to reach puberty is also likely to be a late developer intellectually.

Finally, we must not forget the ways in which the school itself can, and often does, create learning difficulties. The school can create emotional barriers to learning as readily as the home can; it can destroy the individual's motivation to learning and the confidence with which he approaches it; it can offer a social environment that militates against effective learning at the academic level as well as the social; and it can adhere too closely to the kind of curriculum that certain pupils find it difficult to relate to or to find meaning for themselves in (Young 1971).

Withdrawals

One of the more obvious ways in which the school can do this disservice to those of its pupils who have learning difficulties would seem to be by segregating them from their fellows in a permanent way which labels them as different, inferior, difficult, ineducable, nonexaminable or whatever. Nothing can do their social and emotional development more harm and nothing can do their intellectual development less good. It has often been said that if nothing succeeds like success, then conversely nothing fails like failure. We have mentioned many times the psychological and social reasons for not segregating such pupils in this way and for allowing them to work with others on as many common activities as possible. The dividing line between the provision of special treatment and the establishment of a ghetto is very narrow. Pupils who are experiencing learning difficulties should not be singled out for all purposes and they should not be made to feel that they have no valuable and valued contribution to make to the life and work of the school.

If this is in part the thinking behind the move towards a mixed-ability form of organization, it would seem a contradiction of that thinking to use the individual assignment approach as a device for singling out the less able for remedial treatment at a time when the others are pursuing their own individual or group assignments. One of the strongest arguments for this method, as we have seen, is that it allows every pupil to work at his own pace and at his own level, and since, as we have also seen, learning difficulties cannot be generalized and each case is unique, the individual assignment approach would seem to be most appropriate as a means of diagnosing and dealing with each individual's personal educational problems without viewing him as a member of a special group requiring special 'compensatory' or 'remedial' treatment.

Clearly, such pupils will need remedial help in some areas, perhaps especially in basic reading skills, and we must not ignore the development of these important basic skills. Many schools have set out to meet the problems posed by these pupils through a system of withdrawals. At certain times of the week those pupils who need extra help with such basic skills as reading or writing are withdrawn from the class to be given special lessons either by specialist teachers within the school or, in some cases, at a reading centre or other such establishment outside the school. In this way those teachers who have special gifts of this kind or special training can use their talents to the benefit of all the pupils who need this kind of help and not just those who earn a place in a 'remedial' class.

Clearly this kind of help is needed, since too often the teaching of basic skills is not attended to beyond the infant school. However, it is important that it should not be provided at times when these pupils will have to miss what the rest of the class is doing in order to take advantage of the facilities that are offered. For there are a number of problems that are created if this cannot be avoided, such as the risk of a lack of continuity in the work of these pupils in the ordinary class context or, worse, of any group or groups they are attached to, as well as the stigma that must go with any suggestion of being taken out of the normal routine for some special purpose.

For this reason, some schools arrange things so that at certain times the mixed-ability class does not meet as a class but breaks up into a number of different groupings in which all pupils are given some kind of special provision. This is one aspect of that flexibility of grouping we spoke of in Chapter 1.

Most pupils need some kind of remedial help. Some bright pupils will need remedial help even with their reading. Few of us have no learning difficulties at all and these difficulties must be seen in relation to specific areas of work and particular skills, not as evidence of some general disability. Furthermore, at such times pupils can be given additional work of all kinds to suit their needs.

Such a scheme is operated at Northcliffe Community High School, as Bruce Liddington describes (1975, p. 135).

> However, for a certain time, *daily* — for progress will be inhibited if each day does not take up from the one before — such children need assistance in reading and basic Mathematics such as cannot be given by the non-specialist in the full-class situation. At Northcliffe we have a Reading Centre, which is furnished with highly specialised equipment — and carpets and armchairs. Such an attractive and comfortable room acts as an

incentive for the children to be relaxed and responsive and makes it less likely that they will be resentful of going there. It is staffed by a full-time remedial teacher, who not only organises the children but also advises other staff on the content and style of their courses.

A similar scheme is described by Andrew Hunt as operated at Sir Leo Schultz High School (1975, p. 70).

> The school's Remedial ('Opportunity') Department operated on a withdrawal basis, attempting to 'cure' retardation to the point where almost all pupils could be integrated into the school's normal teaching programme with only occasional 'reference back' to the remedial area itself. This work, mainly concerned with basic reading and comprehension, was the province of an experienced teacher who had the status — and salary — of other general Heads of Department in the school, and who was assisted wholly by the intake co-ordinator and in part by other members of staff with special interest and skills in remedial work. Quite simply the Opportunity Department attempted to 'free' itself of its existing 'clients' at the end of the first year of study in the school (i.e. by the 14+ stage). By and large, it succeeded, so that only a handful of pupils with specific and intense learning difficulties were to be found from the 14+ 'options' stage onwards.

It has also been found possible by some schools to link this kind of provision, as was suggested earlier, with arrangements for attention to other special needs. Additional work in some areas for the highly gifted pupil, if this is felt to be necessary, can also be made available at these times. In short, a flexibility of grouping, if allowed for in the timetabling arrangements, can be used to cater for a variety of special needs.

If provision of this kind is properly made, teachers need not feel obliged to try to attend to the teaching of basic skills or the meeting of remedial needs in other contexts. Time allocated for individual and group assignment work should not be seen by the teacher as an opportunity to mount an intensive remedial campaign on his poor readers, except in so far as it gives scope for encouragement and enhanced motivation or provides cues for the teaching of basic skills. To do this is to deprive the other pupils of their share of his attention, to make the poor readers unduly conscious of their difficulties and to deprive them of the opportunities and advantages that individual and group assignments offer to all pupils and to them in particular. For the advantages that have been claimed elsewhere in this book for such work and the advantages of group work in particular are especially important in the education of pupils with learning difficulties.

The special advantages of individual and group assignments

In the first place, it has been argued that when pupils are allowed to work at their own pace and level, and perhaps also in pursuit of their own interests, there are great gains in motivation and incentive to work and to learn. This is precisely the kind of gain we are looking for in the education of the pupil with learning difficulties. If he can be allowed to do what interests him, if he can be encouraged to take on what he can cope with, if we play to his strengths rather than his weaknesses, then we can reasonably hope to achieve a level of motivation that may give us rather more chance of doing something for such a pupil than we have often had in the past. If his attitude towards school and teachers improves as a result of this so that he ceases to offer behavioural difficulties, something of value will have been gained; but we can hope to gain even more if we succeed in building up his confidence in this way. We mentioned in Chapter 2 the stress placed by psychologists such as Piaget on the part played by intrinsic motivation in intellectual development. Allowing the pupil who is experiencing learning difficulties to work from his own interests is one way in which we may hope to develop intrinsic motivation and thus assist his intellectual development.

A second advantage that has been claimed for individual and group assignment work, especially when it is also interest- and enquiry-based, is the opportunities it offers for the promotion of creative work of all kinds. Again, such work offers special advantages to those who are having difficulties in other areas of their work and experiencing the emotional problems that can be associated with such difficulties. It is often claimed that such pupils learn primarily by doing. This is probably true of most pupils but is perhaps particularly important to those with learning difficulties. Drama, movement, art, craft and handwork of all kinds are important to all pupils but their importance to pupils with learning difficulties is vital. Such pupils are probably experiencing the greatest difficulty in expressing themselves through the written word and perhaps the spoken word too; reading and writing difficulties will present them with real problems of communication. Other media can offer them a means of expression denied them through language. Pupils who appear to have very little to show for their years of schooling can often be seen nowadays expressing through dance ideas of a complexity that would have been impossible for them to express through words.

Opportunities for self-expression of this kind will contribute also to their emotional development since their feelings need not remain bottled up inside or manifest themselves in bad behaviour but can be channelled into socially acceptable modes of expression through such art forms. Work of this kind will also promote learning in other fields, since, if satisfaction is achieved, they will want to talk about what they have done and might even be persuaded to write about it; they might also be led by a skilled teacher from such activities into others that arise naturally from them.

Furthermore, these are areas in which the confidence of such pupils can be developed because they can perhaps achieve something worthwhile here. There are those who, although experiencing difficulty with reading and writing, reveal real talent for painting, modelling, dancing or acting. If they have such strengths, the teacher should take advantage of them. Even those who do not possess talent in these directions however, can be helped by an imaginative teacher to achieve real success in these areas. Art does not have to be conceived as painting or sculpting; it can be seen as the creation of a 'montage' out of a variety of materials. Dance does not have to be interpreted in a strict balletic form; it can be viewed as free expressive movement. Drama need not be restricted to the acting of plays written by others; it can be the expression of one's own feelings through mime or improvised dialogue. In short, although all of these areas can provide opportunities for some pupils to reveal and develop great skills and although it should be the teacher's aim to enable pupils to develop these skills to the highest level in order to achieve the greatest scope for self-expression and satisfaction, skilled performance is not always an essential and all pupils can achieve some satisfaction in them.

Nor need we always expect less skilled performances. Again, Bruce Liddington's comments are worth noting (1975, p. 138).

> The school play can also offer a vehicle for the child who in most other respects has little opportunity to stand out from the throng, to be an important segment of a whole — a vital cog in the wheel. Nor need the longest parts necessarily be given to the best readers, with the willing but less able pupils as spear carriers. In spite of initial sight reading difficulties, I have seen children of well below average academic ability grow visibly in stature during rehearsals to overtake their erstwhile superiors on the home straight, once the barrier of learning the words has been overcome.

A third main area in which it has been argued that individual and group assignment work offers advantages to pupils is that of social learning, which, as we have seen, also has its effects on academic learning. Again, this is the kind of advantage that the pupil with learning difficulties most has need of, since, as we have also seen, if his difficulties are not the result of the social climate of his home or the school itself, they will certainly be a potential cause of the development of certain social or anti-social attitudes. His social education must, therefore, be handled very carefully.

We have stressed that the move to mixed-ability groups implies a move away from competition towards cooperation as a prime educational principle. The pupil with learning difficulties needs a cooperative rather than a competitive atmosphere in which to work if anyone does. Competition can only draw attention to his deficiencies and too much of that will cause him to become disheartened. A cooperative atmosphere, on the other hand, can do much to build his confidence by showing him that his work is worthwhile, that it is valued by others as adding something of value to a joint undertaking and is thus as worthy of respect as the work of the more gifted pupils. Only in this way can we hope to promote his social learning and avoid the development of emotional difficulties.

The emotional development of all pupils requires the teacher's careful attention but that of those who are experiencing special difficulties requires particular attention since, as a general rule, their emotional development is most at risk. To have a mental age of twelve at a chronological age of sixteen is not to have an emotional age of twelve. It is, however, to be in danger of developing serious emotional problems. These problems of social and emotional development may be especially important in the final years at school, since it is at this age that the adolescent is becoming socially aware and needs to be enabled and encouraged to make contributions at a number of levels that can be seen to be socially valuable.

It is for this reason that many schools have found it helpful to engage pupils in their final year in various kinds of community service. Such activities are to the advantage of the community but the one who really profits from them is the pupil. They are advantageous to all pupils but particularly so to those with learning difficulties. If teachers are not aware of these needs and do not cater for them, they can and should expect such pupils, especially in their last year of school, to give them hell, as so many of them do.

In general, what is being asserted here is that remedial education, or whatever we wish to call the education of pupils with learning difficulties, is not to be seen merely as a matter of providing additional instruction. Rather, it is a matter of approaching their total education in a different frame of mind, of seeing the need to create the right social and emotional climate for them to work in and of building up their confidence in themselves by playing to their strengths, encouraging them to do the things they can and ensuring that their achievements in these fields are valued by us and by their fellows. Clearly, this requires the mixed-ability class, and within the mixed-ability class it can best be achieved not by singling them out for special provision but by providing for them along with the rest in a situation where each pupil is working as part of a group or subgroup on an assignment suited to his abilities.

However, a weather eye does need to be kept on them. It was suggested when we discussed the grouping of pupils that particular attention would need to be given to pupils with problems of this kind to ensure that they would be accepted by the groups they joined and would have a genuine contribution to make to them.

There are two aspects of this. In the first place, it has just been stressed that the social development of these pupils may be particularly troublesome and teachers must keep their fingers on that pulse. Secondly, a careful watch must be kept on the nature of the contributions they are making to the work of a group to ensure that these contributions really are worthwhile and likely to help towards an improvement in their own learning. Teachers must also remember that this kind of pupil will have more trouble than most in adjusting to the bustling atmosphere of the classroom in which pupils are working on their own assignments. He will have less confidence than most to strike out on his own in this way and will need more support than most in doing so. As we saw in Chapter 5, he will also need more help than most in making the social relationships necessary for working in this way. For all of these reasons the teacher must be particularly conscious of those of his pupils who have serious learning difficulties and of how they are fitting into the pattern of work and of social relationships within his class.

There are also enormous advantages in preparing work-cards at varying levels of complexity to cater for all abilities and in particular to allow for the fact that pupils who have learning difficulties will not be able to progress as rapidly as others through any work that is set for the whole class and to avoid the danger of either boring the quicker workers or overstretching the slower ones. Such a scheme has been devised by those concerned with the development of the work-cards used by the mathematics department of Crown Woods School to which we referred in Chapter 6. Phil Prettyman describes this aspect of these work-cards as follows (1975, pp. 164-165).

> At the moment we have produced self-explanatory teaching material which caters for children of mixed abilities by individualising the pacing. In mixed groups it is usual for only four or five children to experience basic problems with the work so that our system works reasonably well, but there is plenty of room for improvement. The scheme should develop from the strictly linear to contain loops and jumps that will allow the content to be intensified or thinned out for those children who experience specific problems at particular levels, or for those for whom an unnecessarily large amount of time spent dealing with elementary work will lead to the very boredom and carelessness in Mathematics which we were hoping to eliminate. Furthermore, the very able child, of which we have a handful each year, is at the moment being catered for by the basic course, augmented by books and ideas at the disposal of the teacher, but these children's abilities could be harnessed far more productively than we are doing now, and certainly more productively than we have ever done in the past.

The poor reader

So far it has been argued that the individual approach to education that seems to be implied by the move towards a mixed-ability form of organization can cater for the pupil who has learning difficulties as well as any other pupil and that, far from depriving him of the advantages of working in this way, we must see these advantages as having a particular application to his learning difficulties. However, we must now look at the position of the nonreader or the poor reader who would seem to present a teacher with particular problems since he lacks the basic skill to handle work-cards and most resources and it might be asked how he can be expected to take on his own individual assignment or contribute significantly to the work of a group.

In fact, much the same reasoning would seem to apply here as applies in the case of pupils with more general learning difficulties. We must remember that difficulties with reading, like all learning difficulties, can be the result of organic factors, of physical or neurological disabilities, but they are at least as likely to be due to environmental and emotional factors, elements in the pupil's experience at home or at school which have created blocks to his learning, so that even pupils of above-average ability in other areas can be seen experiencing difficulties in learning basic skills such as reading. We saw, when considering pupils with learning difficulties in general, that the solution to those difficulties that arise from factors of this kind is not only to make provision for additional instruction, since that, although important, is to attempt to deal only with the symptoms; it is rather to try also to get at the causes of the disability by creating a secure and stable environment in the school and in the classroom in which the pupil can work and through which his confidence can be built up and by keeping a particularly close watch on his social and emotional development, doing all that we can to make it as smooth as possible.

If this is a correct view of the problem and if it is right to claim social and emotional advantages for individual and group assignment work, then it would seem right and desirable that pupils who are experiencing difficulties in mastering the basic skills of reading should be included in such work. For they, perhaps more than any other pupils, need the advantages that such an approach can bring.

Obviously, as I have already suggested, they will need to be given intensive remedial help with their reading problems, but this should not be done at the expense of the opportunities that can be offered them for working on what they can do and possibly can do well, since to do this might well have the opposite effect to the one intended — it might merely aggravate the social and emotional factors that may lie behind the reading difficulties. If nothing else, these pupils can be given, via an individual assignment suited to their abilities, a rest from the hounding they sometimes get as poor readers from Monday morning to Friday afternoon in a traditional curriculum. For inability to read and write is a disadvantage that is highlighted in almost every curriculum subject. Individual assignment work should be seen as a chance to give them a break from this, to play to what strengths they have, as has already been suggested, and to try in this way to attend to their social and emotional development, which is both important in itself and can have the added advantage of making the way easier for their subsequent intellectual development.

They also need the advantages that can come from the social relationships that develop through working with others on group projects. The practice of attaching poor readers to other groups to work in this way is well established in many junior schools and it is obviously a practice that teachers at secondary level would do well to emulate, not least in the final year where emotional difficulties are likely to erupt in serious behavioural problems.

Working in this way should also lead to gains in incentive and motivation, as has been suggested before, and since inability to read may stem from the absence of any incentive towards acquiring this skill, there is some hope that in certain cases the development of interests through individual assignments may lead to an increased level of motivation towards learning to read. It must be remembered that it is not only books that offer opportunities for reading or require this skill to be understood. Poor readers can be led into acquiring this skill by way of newspapers and magazines and also by way of timetables, road signs, labels on foodstuffs and other household commodities, recipes in cookery books and many other such sources. There is scope for developing an interest in reading by this kind of route in individual and group assignment work. However, these pupils do find reading difficult and, although we may hope to build up their confidence and interest in this way, we must consider how they can be catered for in individual and group assignments.

In the first place, we will need to be particularly careful to develop resource material that they can use. It is obviously vital, therefore, that we think of resources in terms other than those of books, pamphlets and other written materials. It is for the poor reader in particular that we need to develop other kinds of resource. It was suggested, when we were discussing in Chapter 6 the provision of resources, that we should prepare tapes for poor readers instead of work-cards and that we should have ready as many resources that can be used without reading skills as we can obtain or make. The aim of individual and group assignment work is to make the pupil independent of the teacher and, if the teacher is not to spend an unfair proportion of his time with the poor readers and thus neglect those pupils who can read, he must be prepared to help them to become independent by taping material himself and making other provision of such resource material as they can use. Furthermore, this can help in the development of their reading skills too, since they can be given a written version of all or some of what is on the tape and can follow it as the tape is played.

Nor should we assume that printed matter cannot be used by nonreaders. Although we sometimes speak of these pupils as nonreaders, they will in fact be backward readers at various stages in the development of their reading skills. What is needed, therefore, as was suggested in Chapter 6, is the provision of the same resource material at several different levels of complexity to cater for the different levels of reading skill that we will find in any mixed-ability class. If we do this successfully, only the complete nonreader will be unable to cope with the written material we have and he can be catered for by the provision of tapes and other kinds of resource as has just been suggested.

Furthermore, it has already been suggested that such pupils can be employed as illustrators in a group project or can contribute in some other way that does not entail reading or writing. They can also perform a very useful function, valued by their colleagues and giving them a sense of having something important to contribute, if they are encouraged to take responsibility for the use of some of the hardware the group may need. To operate a film-strip or slide projector, a tape-recorder or a camera requires skill of a different kind and this is the kind of contribution such children can initially be encouraged to make. The attitude of some pupils to school and to teachers can be transformed if they are given this kind of responsibility. One pupil in particular, I remember, a fifteen-year-old in his last year at school, who was a serious threat to the safety of teacher and pupils alike, until the tape-recorder was put into his care and he was given responsibility for all recording. Such pupils need to feel valued, as we all do.

We must beware, however, of appearing to suggest that such pupils should be used in group projects as handmaidens or attendants, assisting in the development of the ideas and the work of other, brighter pupils. As was stressed earlier in this chapter, our basic educational goals and principles should be the same for all pupils and we must not willingly accept different aims for our teaching of pupils with any sort of learning difficulty. This is one reason why it is necessary to prepare similar resources at different levels of complexity, to ensure that they can be given similar work to do. It may be expedient in many cases to devise useful jobs for them of the kind we have just discussed, in order to secure their acceptance by a group, to build their confidence or even to keep them busy and ensure their good behaviour, but we must never let this cause us to lose sight of the fact that in the end our real concern should be with their education and whatever devices we employ they should always be seen as subservient to this end which is common to all pupils.

A third major consideration for the teacher, which again applies to all pupils but perhaps particularly to those experiencing difficulties with reading, is the importance of providing opportunities for the development of spoken language. As the Newsom Report told us (§467), 'Inability to speak fluently is a worse handicap than inability to read or write. . . . Personal and social adequacy depend on being articulate.' We might go further and claim that the development of the pupil's skill with spoken language will also lead to an improvement in his control over the written word. Children's language skills do not improve merely by the performance of written exercises. They improve as much, and in the early stages of language development considerably more, through opportunities to talk to others about what they have done and about what they are planning to do and how they intend to set about it.

Language will develop by its use in expressing ideas that are important to the learner and will develop, therefore, by being used to describe activities and interests the pupil is engaged on. It can, therefore, be promoted by the kinds of activity associated with individual and group assignments, if teachers are conscious of its importance and prepared to provide opportunities for it. Formal reporting back to the class of progress made and of future plans has its place here and one should not discourage the poor reader from acting as spokesman for a group if he wishes to, but opportunities for informal discussion are probably of more value to all pupils and especially to those with reading problems. It is in this kind of area that the mixed-ability class and the mixed-ability group within the class offers particular advantages, for there can be little possibility for development of language skills if all members of the group are at roughly the same level of linguistic competence and proficiency. It is in a mixed-ability group, where the pupil who can express himself more easily must work to communicate with his less fluent fellows and they in turn can develop by their contact with those more fluent than themselves, that improvement can be hoped for in every pupil's linguistic ability.

Again, however, the teacher has to be aware of the possibilities here and work deliberately at taking advantage of them in ways such as Shirley Legon describes (1975, p. 143).

> During discussions I make a point of catching the attention of the slower ones and communicating with them silently: a querying look (do you understand/agree/are you all right?), or I raise my eyebrows incredulously at them (doesn't she go *on*?). If it looks as if they have some-

thing to say but are not very articulate I remark that other people like Ann, for example, would like to get a word in, sometime, and invite the contribution, helping out if necessary. It is a struggle sometimes to make them listen patiently to each other and not interrupt or call out derisively, but we do succeed as time goes on.

Further ways in which an incentive towards learning to read might be created and a basis provided for some written and oral work may be found through the creative activities discussed elsewhere and the kind of community service work recommended earlier as suitable for pupils with learning difficulties, particularly in their last years at school. Both of these things, however, have an equal value for all pupils and, although their value as starting-points for reading and other learning may be greater for those who have learning problems, we must again be wary of appearing to suggest that these opportunities should be provided only for such pupils. We will need to discuss these points more fully when we come to consider the final years of schooling in Chapter 9.

A final word must be said about the correction of such written work as the poor reader may be encouraged to produce by the teacher's efforts in the individual and group assignment situation. He must not be dis-heartened and discouraged by having every error indicated. After all, we will have had to work hard to build up his confidence and interest to the point where he has produced this work; we do not want to destroy the fruits of that by a too liberal use of a red pencil. Such work must be seen as a growth point, as an opportunity to lead the pupil on further, and we should, therefore, only draw his attention to errors we feel can now be corrected as the next stage of that growth. We should mark to encourage not to discourage and should bear in mind the development of the individual's ability to communicate rather than to spell or punctuate.

Summary and conclusions

In this chapter I have tried to list some of the reasons why many pupils might be experiencing difficulties with some or all aspects of their learning. It has been suggested that for such pupils, perhaps more than for any others, there are enormous advantages in the mixed-ability class, not least those that derive from the opportunities that working with their brighter peers can give both for their own learning and for the development of that confidence that is essential for any kind of success.

In fact, I have argued that, although it has often been claimed that such pupils have most to gain from a system of streaming which segregates them from the more successful pupils, the opposite is rather the case. It was further argued that such remedial help as they need should be provided by a system of withdrawals and that this system should be combined with an arrangement to withdraw all pupils for some special purposes at set times in the week. We then considered some of the practical problems that the presence of such pupils, and especially those experiencing difficulties with reading, create for teachers and suggested some of the ways in which these problems might be met.

Pupils with learning difficulties must be seen as an integral part of a mixed-ability class and not as a special case requiring separate treatment. If this is so, all teachers need to be prepared both to understand and to cater for such pupils. All teachers employed in infant and junior schools need to be prepared to teach reading as a basic skill. Too often it is the infant teachers only who can do this and we forget that many pupils still need to be helped in this way in the junior school. At secondary level, however, it is probably wiser to leave the teaching of reading to teachers who have been specially trained to attend to this in the special classes referred to earlier. Not the least important reason for suggesting this is that teaching of such basic skills at this age is essentially remedial teaching, since by definition every child who needs this kind of help at this stage of his schooling is experiencing difficulties of a very special kind with all the possible emotional side-effects we have noted.

This does not mean, however, that secondary teachers in general can continue to ignore the problems of learning to read and leave them to the specialist. An important distinction has to be made between the actual teaching of the basic skills of reading and making due allowance for the nature of reading and language development generally in all the work we do. Reading, as part of language development, has to be seen as an 'across the curriculum' concern. All teachers, therefore, will need an understanding of the nature of reading and the nature of the reading difficulties children suffer from. Even the subject specialist needs this kind of understanding if he is to provide his pupils with the kind of reading material they can profit from. This will be especially important in his teaching of mixed-ability classes since he must be able to choose a variety of reading materials suited to the needs of different individuals or groups across the whole range of ability.

Furthermore, it is apparent that some pupils experience reading or language difficulties in certain subjects only, sometimes because of the peculiarities of the language that teachers, for the best of reasons, insist should be used in certain subjects or disciplines (Barnes 1969), and this puts an onus on the teacher of a subject to acquire this kind of understanding of the nature of reading and language development generally in order to be able to understand the kinds of problem facing certain of his own pupils in his own subject area.

Finally, few teachers would deny that most of the difficulties they have with control of their classes and the behaviour of certain pupils stems from those pupils who are experiencing serious learning difficulties. In a streamed school the 'C' stream has always been more difficult to handle than the 'A' stream and '4C' has been the death of many a teacher. Some of the reasons for this we have mentioned several times already. The move to mixed-ability classes can be seen, therefore, in terms of a philosophy of 'divide and rule'; it should be seen rather as an attempt to get to the roots of this difficulty and cure it by providing these pupils with opportunities to do something valuable and constructive alongside their fellows. If teachers do not succeed in helping them to take advantage of these opportunities, the behavioural problems will continue and the change will have been largely wasted.

CHAPTER 8

TEACHER-PUPIL RELATIONSHIPS

'When those you rule are unruly, look to your rule.' This, if not one of the more amusing, is certainly among the more significant of the sayings attributed to Confucius. Although there are pupils whom Confucius himself would find it difficult to handle, most problems of control in schools arise from the teacher-pupil relationships that the school and its individual teachers generate and these in turn depend to a considerable extent on the methods of control and the patterns of organization adopted. It would seem reasonable to assume, then, that a change in the organization of the school such as the introduction of mixed-ability classes will have major implications for the discipline within the school.

Furthermore, reference has already been made several times to the growing body of evidence of the effects of a streamed form of school organization on the behaviour of pupils (Hargreaves 1967, 1972; Lacey 1970), so that it is not unreasonable to assume that the introduction of a mixed-ability form of grouping should lead to a corresponding improvement in the disciplinary situation and an alleviation of problems of control.

The advantages of mixed-ability grouping

It would be a mistake, of course, to assume that all the school's and the teacher's problems are solved as soon as streaming is abandoned and mixed-ability classes are established. In the earlier chapters of this book it has been the aim to make it clear to teachers that they must work hard to make this change work as well as to suggest some of the ways in which they may set about this task. Mixed-ability classes provide teachers with greater scope and with improved opportunities but the onus is on the teacher to take advantage of the opportunities offered. Similarly, it would be a mistake to assume that with the introduction of mixed-ability classes all problems of discipline and control disappear as if by magic. Indeed one can see that in one sense this kind of change creates more scope for misbehaviour and makes the teacher's task of controlling his class more difficult, since if pupils are to spend a lot of time working independently, they will need to talk and to move about the classroom and perhaps even the school building with some freedom and will of necessity be left much more to themselves, to work or not to work, than in a traditional class-teaching situation. Potentially, then, there would seem to be dangers here that the teacher should be aware of.

On the other hand, there is some evidence to suggest that a move to a mixed-ability form of organization is more usually accompanied by noticeable improvements in behaviour and a reduction in the problems of discipline and control (Kaye and Rogers 1968). It has been noted that pupils' attitudes to school become more positive (Barker-Lunn 1970), that many of them cease to be difficult and that this improvement permeates all aspects of their work. In schools where they have been given a limited opportunity to work in mixed-ability situations, pursuing their own interests and investigations for only a part of the working week, improvements have been detected in their attitudes to work not only in that situation but also in the more conventional lessons they continued to have at other times. Typical of the comments made by teachers and headteachers on this phenomenon are those of Arthur Young, headmaster of Northcliffe Community High School (1975, pp. 32-33).

> I suppose it took about six months for the effect of the first moderate move towards unstreaming — the broad banding — to become apparent. The pattern of behaviour became more even — approximating more to the 'A' and 'B' streams that we used to have than to the former 'C's' and 'D's'. There were still some classes which behaved better than others — there

always will be, I suppose — but the clear distinction between the goodies and baddies seemed to be disappearing. We were also retaining more of our Duke of Edinburgh Awards candidates, although we still had a majority coming from the more able band than from the other band. We discussed the results of our changes at monthly staff meetings, and found general agreement that we had improved the morale of the boys appreciably; the standards of work were not suffering and many of them were making more of an effort to show us that our trust was not misplaced.

Again we must stress, however, that this is not an inevitable concomitant of the change; it is the result of the way in which teachers use the advantages the change gives them to develop more productive relationships with their pupils. That there are such advantages there is no doubt and to use them it is necessary to understand the form they take.

We must first of all take note of the fact that the move to a mixed-ability form of organization removes certain factors that seem to be productive of behavioural problems in the streamed situation. It was said in Chapter 7 that pupils with learning difficulties, when grouped together in a 'C' stream or a remedial class, can present the teacher with more discipline problems than most classes. Few teachers can fail to be aware of this and all will know only too well of the difficulties such classes can create in their final year of schooling. Nor should any teacher be surprised that pupils who have always experienced failure at school should want to fight the system and their teachers as representatives of it. As we have seen, such pupils, finding little or no satisfaction in what the school is offering them but being required nevertheless by law to present themselves each day, will seek satisfaction in their own affairs, from their own kind and through the things that interest them. They will develop their own subculture within the school and this will be of a delinquescent and disruptive kind (Hargreaves 1967, 1972; Lacey 1970).

Unstreaming will ensure that such pupils are not gathered in what have been called 'sink' classes and will thus avoid a clear labelling of them as failures. In this way it will remove one of the prime sources of behavioural problems and provide the teacher with a situation in which his attempts to establish good relationships with such pupils are not doomed to failure from the outset.

A second source of behavioural problems which is removed by the introduction of unstreaming, or at least by the individualized learning with which I have argued it should be associated, is the alienation of many pupils from the content of their learning.

The concept of alienation has a long and complex history spanning the last two centuries, but basically it is a term whose use implies that a man-made creation achieves a pseudo-life of its own and begins to control man who created it. Thus Feuerbach, Marx and others have described man as alienated in relation to religion, government, work and so on and have seen this as an unhappy and 'unnatural' condition productive of friction and in some cases ultimately of revolution.

It is not uncommon for pupils to experience the same kind of alienation in relation to the traditional school curriculum or any curriculum that is totally teacher-directed. For a gap is created between their out-of-school interests and what they are required to involve themselves in inside the school, between the values exemplified in their own interests and those of their family and friends and the values implicit in the curriculum imposed by the school (Young 1971). We might express this differently by using the terminology of A. N. Whitehead (1929) and speaking of the knowledge they are being required to assimilate as a body of 'inert ideas' which in no way becomes a part of them but remains for ever separate, acquired, inactive and without real meaning for them.

Again, this will be more commonly the experience of the pupil with learning difficulties; it will also be the experience of the pupil who comes from the kind of home in which the cultural background is very different from that implicit in the school curriculum; but to a greater or less degree it will be the experience of all pupils presented with a curriculum that comes entirely from outside them. It is also a phenomenon that is easily recognizable as a source of behavioural problems. As such, it is largely removed by the introduction of individual and group assignment methods, since even where these are mainly teacher-directed there is some attempt to relate them to the needs of individual pupils, a greater element of pupil involvement and a higher level of motivation.

A further advantage of pupil-choice in relation to discipline and control is that it removes some of the need for compulsion that is so often a source of difficulty in dealing with children, and particularly with adolescents. Nobody likes to be forced to do anything: to be forced is to be deprived of the freedom we all value and, while one would not want to claim that compulsion is never necessary, it is to be avoided whenever possible in the interests of harmonious relationships. Compulsion is also at variance with the individual approach we have argued is essential in the teaching of mixed-ability classes and for this reason too is unlikely to lead to the successful handling of such groups of pupils. If we are prepared to trust pupils with responsibility for their own work, their own learning, their own education, not only are they likely to respond to such trust, we are also creating a climate in which the head-on clash that is the result of attempts at direct compulsion and which is seldom productive of anything but friction, often accompanied by a loss of face for the teacher, can be readily avoided and a basis for a much more constructive kind of authority structure established, as we shall see in a moment. If we can do this, then a very real source of behavioural difficulties will be removed at a stroke.

A fourth potential source of discipline troubles is removed by the change of emphasis from competition to cooperation which we have seen is implied in the introduction of a mixed-ability form of organization. Competition must generate friction and must lead to problems of behaviour in all pupils but especially in those who are not succeeding in the competitive atmosphere of the school or class. If we discourage competition, then, we remove another source of potential trouble.

At the same time, the introduction of a more cooperative approach to learning brings with it further advantages to the teacher in the matter of control and leads us on to a consideration of more positive aids to discipline that exist in the mixed-ability situation. For the move to mixed-ability classes not only removes certain factors from the teaching situation that have been productive of unsatisfactory relationships between teachers and pupils, it also introduces into the situation certain factors which, if properly used by teachers, can lead to much improved relationships. The move towards cooperation is one of these. For the mixed-ability class brings with it the notion of cooperation at all levels. Certainly one basic aim is to encourage a more collaborative approach to learning among pupils, so that relationships between pupils should be improved and the kind of friction that leads to behavioural troubles reduced.

However, such an improvement in pupil-pupil relationships can only come about if there is a corresponding improvement in all relationships. We saw in Chapter 4 that team-teaching with its emphasis on teachers working collaboratively with each other is an almost inevitable development from the introduction of mixed-ability classes, so that improved relationships between teachers also often follows from unstreaming. Many would also wish to bring parents more fully into the process and would wish to work for improved teacher-parent and child-parent relationships. Even if one does not go this far, there is no doubt that different pupil-teacher relationships are necessary, since a collaborative atmosphere can only be engendered if teachers collaborate with pupils as well as encouraging pupils to collaborate with each other. If this kind of improvement in teacher-pupil relationships can be brought about — and the mixed-ability class makes it possible as well as necessary — then a positive step will have been taken towards the removal of many potential discipline problems.

A second major factor in the mixed-ability class that can make a positive contribution towards improved teacher-pupil relationships and, as a result, an improved state of order within the classroom arises from this general point about collaboration. Attention has been drawn several times to the importance of creative work in the mixed-ability class and to the increased emphasis that should be placed on this kind of work. Such work both requires and contributes to differently based relationships between teachers and pupils. We can expect little creative work from pupils in a strict, authoritarian situation since this is likely to inhibit rather than promote the freedom of expression that creative work requires. A much more relaxed atmosphere is needed if pupils are to feel free to try things out. They will need to be confident that the teacher will look sympathetically on their failures as well as approvingly on their successes. Such work, then, requires teacher-pupil relationships that are based on collaboration and mutual understanding rather than on distance and control. In turn, these activities will promote this kind of teacher-pupil relationship since it will be through them that this kind of relationship based on mutual confidence will develop. There is nothing like mutual participation in a worthwhile exercise — a play, a film, a dance, a piece of music — to promote the kind of relationship we are discussing. Here, again, the mixed-ability class offers the teacher the opportunity to solve his problems of control by improving the quality of his relationships with his pupils.

One of the reasons why this kind of emphasis has been placed on the provision of opportunities for creative work in the mixed-ability class is the contribution that it can make to the emotional and social development of pupils. This concern with social and emotional development can itself be a third source of advantage to the teacher in the matter of the development of relationships with his pupils. For the teacher who is aware of his responsibility for this aspect of his pupils' development and concerned to fulfil that responsibility will quickly realize that he can only do so satisfactorily if the relationships he promotes are suitable. As we saw in Chapter 5, social learning is caught rather than taught and the kind of social learning that goes on in his classroom will depend largely on the kinds of relationship he is able to develop with his pupils. Again, however, at the same time as this creates a task for the teacher, it also offers him an avenue towards its achievement, since the very display of a concern with the emotional and social development of pupils will contribute to the creation of the kinds of relationship conducive to it. One cannot be genuinely concerned with a pupil's welfare without eliciting some kind of positive response from that pupil.

All of these factors, then, which are present in the mixed-ability organization create a situation in which it is easier for a teacher to develop good, productive relationships with his pupils and to avoid the worst disciplinary problems. To do so, however, he must be aware of these factors and must take advantage of the opportunities they offer.

It will now be apparent, however, that the relationships they will help him to develop and, indeed, the kinds of relationship that they make essential are rather different from those that exist where a more formal approach to teaching is adopted and a more limited set of educational goals. We must now consider some of the characteristics of these new kinds of relationship.

New kinds of relationship

In the first place, it will be apparent that such relationships will need to be at a very personal level. We have stressed the need for individualized learning in the mixed-ability class and for the teacher to be conscious of his responsibility for the education of thirty or more individuals rather than of one class. It will be necessary for him to get to know and understand his pupils as individuals, therefore, and this can only be done if his contact with them is a really personal one. It will also be necessary for him to make the appropriate provision for them as individuals and to cater for all aspects of their development — social and emotional as well as academic — and this too will require personal contact with them. The class lesson, which can be a rather impersonal affair, is not the best aid to education with a mixed-ability class, so that, whether he likes it or not, the teacher cannot do his job effectively by remaining at a distance from his pupils; he must work with them, rub shoulders with them and develop relationships with them that are more personal than those that are necessary and possible in the class-teaching situation.

A second, and perhaps more fundamental, feature of these new kinds of relationship is that they are the result of a concern with education in its fullest sense. If we are right to claim, as we did for example in Chapter 2, that education is concerned with the development of the individual's ability to think for himself and the promotion of individual autonomy, then this cannot be promoted by authoritarian methods of control but requires the kinds of teacher-pupil relationship by which the development of the pupil's autonomy, personal responsibility and independence can be assisted. If education implies the kind of freedom that one associates with a democratic way of life, then it will require a democratic atmosphere for its development. Reference has already been made in Chapter 5 to the apparent advantages of a democratic organization for academic learning; such an organization is essential if education in the full sense is to take place. We cannot educate by coercive methods. You can drive a horse to the water, but you cannot make him drink. If you want him to drink, you must show him that it is a good thing to do so. If you want a pupil to accept the education he is offered, you must persuade him of its value. Our methods should be normative, therefore, rather than coercive, since, as we saw when discussing the value of intrinsic motivation, it is part of what it means to be educated to be brought to see the value of one's education. Nor is it enough to introduce this element in the last year or two of schooling. By then it is too late — especially for the less able. Pupils need to have experienced this kind of approach throughout their schooling, as those schools that have tried to introduce this kind of change only with their early leavers have discovered. A democratic atmosphere is essential to education at all stages and this implies teacher-pupil relationships that are in some sense democratic.

This does not mean that the teacher enters such relationships from a position of equality with his pupils or that he abandons all responsibility for the making of decisions. To say that he should not be authoritarian is not to say that he should not exercise authority; it is to say that he must look very carefully at the nature of the authority he is exercising, since its basis will be different in a democratic setting from that which it has been in an overtly autocratic one.

The teacher's authority is derived from a number of sources. He is a traditional authority figure in so far as he holds a position traditionally associated with the giving of instructions; he has the legal backing — whatever it is worth in practice — of those people and bodies who gave him his position; he should have the personal qualities necessary to secure the acceptance of his authority; and these should be associated with an expertise relevant to his job as a teacher. His authority, then, will be based both on his position and on his expertise. To put it differently, he will be *in* authority and he should also be *an* authority (Peters 1966).

The changes that are taking place in our schools and classrooms and the changes in teacher-pupil relationships we are considering here can be summarized as the results of a move away from the positional sources of authority towards the expert, a reduction of emphasis on the teacher as *in* authority and a consequent increase of emphasis on his position as *an* authority. This reflects changes that are taking place in society as a whole, changes that are themselves the results and the overt manifestations of the move from an authoritarian to a democratic way of life. Less and less are we willing to accept authority based on position; more and more do we want to be shown that the individual has that right to exercise authority that derives from an expertise, a superior knowledge in the field concerned. It is precisely the same in the modern classroom. The teacher whose authority will be accepted is the one who makes it clear that it is based on expertise, not the one who expects to be obeyed simply because he is a teacher. The new relationships to which we have been referring are those that the teacher is able to develop when his authority is accepted in this way.

There are two aspects of this expertise of the teacher, two areas in which he must be *an* authority. In the first place, he must be an authority in some field or fields of knowledge, since he must be able to meet the pupils' demands on his knowledge in the area or areas of work he is responsible for — this is one reason why team-teaching is desirable, so that the individual teacher does not have to take responsibility for more areas than he is expert in.

This expertise in itself, however, will not take him far, partly because his knowledge may not be of a kind that his pupils see as worth acquiring for themselves and partly because it will in any case provide him with authority mainly in one narrow area, although there would of course be some 'spin-off' into other areas. It is for this reason that the teacher needs to be an authority in a second major sphere, that of education. He must show a professional expertise as an educator if his authority is to be accepted. It must be apparent that he knows what is best for his pupils' continued development and education and that he has the knowledge and the skills to help them to achieve it. This is the only kind of expert authority that will enable him to handle pupils with learning difficulties and others who are not motivated by a great desire to share his knowledge of history, physics, or some other subject; it is the only kind of authority that will be accepted by pupils in their last year at school who are also usually unattracted by expertise in school subjects; it is the only kind of authority that will work in the mixed-ability class where, as we have seen, there are many concerns beyond intellectual achievement. It is the only kind of authority, therefore, that can provide the necessary basis for the development of the kind of teacher-pupil relationships we have described as necessary if one is to take full advantage of the opportunities the mixed-ability class offers.

To handle a class in this way, to create for oneself this kind of authority, and to develop relationships with one's pupils of the kind described here is not easy. This is probably the most fundamental way in which teaching has become so much more difficult a job in recent years, although as always it has become proportionately more satisfying for the successful. It is particularly difficult for new entrants to the profession and for those nearing retirement, some of whom may feel they are too old to learn new tricks of this kind. Nevertheless, one must see this kind of change as inevitable if we are to offer pupils an education in the full sense of the word and prepare them for a free society. We must finally turn, therefore, to a consideration of some of the ways in which the teacher can achieve authority and establish relationships of this kind.

In the first place, we must stress again the improvement that follows the move to mixed-ability classes itself and the attempts to fit our educational provision to the individual. As we have already said, this will not in itself solve the teacher's problem but it will create a situation in which a solution is possible and it will start him on the road to that solution. For it represents an acceptance of all pupils and a genuine attempt to involve all equally in the life of the school. Seldom does it fail to evoke a quick response. There is a further gain when this kind of move is associated also with some form of team-teaching, since problems of control assume less significance when they are problems for the team rather than for the individual.

An important feature of this phenomenon is that it represents an attempt on the part of the school as a whole to establish a particular kind of ethos. There is very little that an individual teacher can do to change the pattern of authority even in his own classroom if the pattern prevalent in the school as a whole runs counter to what he is endeavouring to create. However, when there is a general move in this direction, as there must be in any serious attempt to establish mixed-ability classes, advantages immediately offer themselves to all teachers. Again, this is well summed up by Arthur Young (1975, pp. 28-29).

> Relationships within the cosmos of the school are all important — they are based on democratic procedures and controls rather than on any system of sanctions or punishments. The informality and natural rapport which arises from an absence of punishment and prefectorial systems is rewarding and rarely abused. The freedom of our boys and girls to dress as they choose and style their hair as they think best is all part of the same philosophy of education and, as this is a case study, must be taken into account. Hardly a traditional establishment, is it? But we would claim that it suits the needs of the locality it serves better this way. We would not claim this to be the only way nor would we expect general agreement with our methods, but in spite of all the apparent relaxation and freedom we give the boys and girls, our end results in public examinations are as good as they ever were before, and the incidence of aggressive behaviour has been greatly reduced.

Andrew Hunt makes much the same kind of case when describing the pattern of staff-student relationships which emerged at Sir Leo Schultz High School (1975, p. 69).

> Difficult to describe adequately, this was based on the well-established principles of mutual respect, tolerance and understanding, but it went deeper than this, largely, I believe, because almost 100 per cent of our pupils did realise that our total educational concern for them was absolutely and genuinely fundamental to the whole 'way of life' in the school. Once an atmosphere of this sort is engendered its effect in a truly educational sense is immense and I believe that it was this sort of 'ethos' which, above all, created the conditions we needed to press forward with our mixed ability teaching programmes.

Of crucial importance, of course, is the attitude of the teacher or teachers involved. It is this that matters to pupils rather than the mechanics of the school organization. It is the attitude of rejection that is implicit in so much of the treatment some pupils receive at school rather than any particular method of grouping them or teaching them that creates the resentment that leads to poor relationships and behavioural troubles. A change to mixed-ability classes, or even to individual and group assignments will not, therefore, achieve much in itself; it must be accompanied by a corresponding change in the attitudes of the teachers. Perhaps the most significant finding of the research that has been done on the relative merits of streaming and unstreaming is the clear indication that unstreaming only leads to improvements in academic attainment as well as social and emotional development when the teachers believe in it (Barker-Lunn 1970).

If the teacher does not believe in unstreaming, he should not teach in an unstreamed school, since he will find it difficult, if not impossible, to achieve anything there. On the other hand, if he is in sympathy with the values implicit in the change to mixed-ability clases and maintains the spirit of that change in his own teaching, then he will already have taken a major step towards the development of the kinds of relationship that are necessary. If his attitudes are appropriate, then, and those of the school as reflected in its organization are suitable, a lot will already have been achieved towards the development of the kinds of teacher-pupil relationship that I have described as necessary and the avoidance of the kinds of discipline problem that we have heard only too much about in recent years. From this point on, success or failure will depend on the individual's own qualities, his ability as a teacher, his expertise.

Clearly, in the kind of situation we are discussing the teacher's personal qualities will play a major part. The freer, more relaxed atmosphere of the classroom in which pupils are working on individual and group assignments makes more demands on the teacher's qualities as a person than the set performance at the front of a relatively homogeneous class. He needs to be able to make relationships on the basis of an easy and confident authority; he must not be heavy-handed; he must be open and friendly; and he needs a sense of humour. Relationships cannot be made from a distance not can they be made by someone who stands too much on his dignity. It must not be thought, however, that he must be a 'born teacher'. I have never been very sure what such a creature would look like and I find it difficult to believe that anyone could be born with the sort of expertise required by a modern teacher. For, apart from the kind of personal qualities that make it possible for him to make relationships with his pupils, the teacher needs a great deal of skill and expertise of the kind that can only be acquired by hard work.

The mixed-ability class makes much greater demands on the teacher's skills and abilities as a teacher and, as we have seen, his authority will derive largely from this source. He needs as wide a knowledge as possible of the subject-matter his pupils will be working on. He also needs the specific skills that have been discussed in the earlier chapters of this book. He needs to be able to work as a member of a team; ne needs to be able to group his pupils smoothly and efficiently in such a way as to avoid the behavioural problems that can arise when this is not done properly; he needs to be able to acquire or create and make available the resources that are needed by his pupils and to forsee their needs so as to have the right material to hand when he needs it and thus to avoid having pupils with nothing to do; he needs the kinds of skill necessary to care for pupils with learning difficulties and integrate them into the work of the class as a whole; in short, he needs to be an expert at his job.

These skills, however, cannot be acquired in isolation, as one might acquire a skill at plumbing joints or laying tiles. They are skills that can only be acquired and effectively employed if they are associated with a deep understanding of the nature of the educational process, the pupils we are trying to educate and the society in which the whole process takes place. An understanding of education is needed if these skills are to be used to any advantage. The teacher's professional expertise, therefore, must be seen as a number of specific skills based on as full a theoretical understanding of his task as he can acquire.

Summary and conclusions

I have tried in this chapter to list some of the advantages that the mixed-ability class offers the teacher in the task of developing more productive relationships with his pupils. It has been stressed, however, that these relationships will be of a different kind from those that appear to suffice in a traditional classroom situation and that they will be such as to make greater demands on his skills and qualities. They make it necessary for him to be *an* authority rather than to try to depend on the fact that he has been placed *in* authority and require of him that he can clearly demonstrate that he has something of value to offer his pupils. They thus place a greater onus on the development of a professional expertise that is based both on a range of pedagogic skills and on a depth of understanding of the educational process.

If he has this expertise, his pupils will recognize it and will respond to it. His authority will be established. He will besides have the skills necessary to avoid problems of discipline and control both with his class as a whole and with individual members of it. For these skills will enable him to organize the work of his classes with an efficiency that will forestall most problems of control. In matters of class control, prevention is not only better than cure, it is itself the only cure. For the worst cases of indiscipline cannot be cured; they can only be prevented. Prevention can only be achieved by employing the skills we have been discussing and in particular by developing the kinds of teacher-pupil relationship that we have seen are fundamental to education in the fullest sense. Control must be seen as a matter of developing relationships rather than of applying techniques.

Much can be done to provide the teacher with this kind of expertise in his initial training if this is approached in the right way. A lot must be left, however, to experience and the teacher must be so educated himself that he is able to profit from his experience, since the experience itself will teach him only what he is capable of learning from it. He must be prepared, therefore, to continue his own education by both formal and informal methods to ensure that at no stage is his professional expertise found wanting. For when it is, his control will have gone.

CHAPTER 9

ASSESSMENT AND THE FINAL YEARS OF COMPULSORY SCHOOLING

Changes in the provision made for the assessment of the work of pupils in secondary schools and elsewhere have been as dramatic in the last twenty years or so as any that have taken place in education. Furthermore, the public examination system is still in a considerable state of flux. Alongside the GCE, itself only established in 1951, has been introduced the Certificate of Secondary Education. With the CSE has come a greater sensitivity to variations of approach and syllabus to be found in different schools and different parts of the country and a growing awareness of the need to devise an examination structure that could be adapted to these variations through individual assignments, assessment of course work and other new examining techniques. This has led in turn to similar adaptations in the schemes of examination of some GCE Boards (Macintosh 1971) and recently to proposals to combine what is best in the CSE and the GCE at ordinary level into a common system of examining at 16+ (Schools Council 1971).

The pattern of examining at 18+ has also been under review. In particular, there has been extensive discussion of the proposal to replace the existing GCE (A)dvanced level examination with a two-tier examination at (N)ormal and (F)urther levels (Schools Council 1977).

Proposals have also been made for the establishment of a public examination for that growing number of pupils who are choosing to remain at school beyond the statutory leaving age although without the intention of following a full two-year 'A' level course. Examinations have already been held for this Certificate of Extended Education and again this is leading to a number of interesting developments such as the linking together of examination boards and the introduction of new examination 'subjects'.

All of this activity indicates the importance of the public examination for pupils, teacher, and parents. The public examination is important for pupils and parents because for them it is the key to careers of various kinds and can provide the entrée into further and higher education. It provides young people with a qualification of national currency that enables prospective employers and others to measure at least some aspects of their attainment against those of others applying for the same kind of post or further course of training or education.

For teachers it is important both because it is valued by pupils and because it provides them with a clearly definable goal to work towards. Teachers want to do what is best for their pupils and they are aware that one of the ways in which they can most succeed in doing this is to help them to achieve the best qualifications they are capable of. They are also aware that the public examination can provide pupils with a source of motivation that will get them down to work and will obviate some of the behavioural difficulties associated with pupils whose sights are not fixed on this kind of external goal. Not the least important reason for the difficulties many teachers experience with classes of early leavers is the fact that they lack any incentive of this kind. I have argued elsewhere in this book the values of intrinsic motivation, of working at something because it is worth doing in itself and have offered this as a strong reason for encouraging pupils to work from their own interests. Furthermore, in Chapter 8 I made out a case for solving some of our behavioural problems not by employing coercive measures but by attempting to use normative means, to show pupils the value of what we are trying to involve them in. Nevertheless, it would be hopelessly unrealistic to attempt to suggest that this is easy to achieve, or to deny that in most situations calculative measures are the ones that work and that the best way of motivating our pupils is to dangle the carrot of public achievement and future advancement before their eyes. Nor must we lose sight of the fact that, as teachers, we ourselves need some goal for our efforts

172 Mixed-Ability Grouping

and are attracted by the notion of some objective assessment of the effectiveness of our labours. Teaching offers little proof of achievement that one can put one's finger on and the public examination can compensate a little for this.

For all of these reasons, then, although some teachers will claim that examinations restrict their freedom and cramp the curriculum, most will admit the key role they play in secondary education. Indeed, it was largely pressure from teachers themselves that brought about the introduction of the CSE for pupils in secondary-modern schools. Active concern with examining procedures stems in part, then, from the importance placed on public examinations by pupils, teachers and parents.

A second reason for the growing concern with the system of public examinations is the implications for that system of the changes that have been taking place in secondary education generally. The move towards the total comprehensivisation of secondary education is having the effect of opening up new educational opportunities to a wider range of pupils and the introduction of a statutory school leaving age of sixteen has made the public examination more important than ever. The strongest argument for this measure is that it enables pupils who before were forced by economic and other circumstances to leave school at fifteen to stay on to an age when they can achieve a worthwhile qualification and this is resulting in an increase in the number of candidates for public examinations. The demand for an examination such as the CEE to be taken after one year of sixth-form work is also the result of an increase in the number of pupils choosing a further 'extra year' at school without wishing to commit themselves to an academic two-year course to 'A' level qualifications. Teachers are also looking to a changed provision to provide an incentive for more pupils, since many are greatly concerned about the potential behavioural problems of the extra year.

The greatest source of pressure for reappraisal of the examining system, however, is the dramatic changes that are taking place in the curriculum of the secondary school and the new attitudes, approaches and methodology reflected in those changes. Recent developments in the form of examinations at both CSE and GCE ordinary level are the first attempts to bring the examining system up to date with the curriculum changes that have taken place and are taking place.

That examinations can and do have an effect, usually of an inhibiting kind, on curriculum development, however, must be recognized, so that, if curriculum change of any kind is to occur, consideration must be given to the corresponding changes that must take place in examination procedures both to allow for such change and to assess the new kinds of work that will be introduced. New patterns of organization and new approaches to the curriculum require new techniques of assessment and a new structure of examinations. It is for this reason that concern about public assessment procedures looms very large in all discussions of the implications of a change such as the introduction of mixed-ability classes.

The introduction of mixed-ability classes, therefore, and in particular the adoption of individual and group assignment methods within these classes will have serious implications for the system of public examinations. It may even be the case that a system of public examinations is incompatible with the individual assignment method of teaching or with the mixed-ability class, and that, whether we like it or not, we may have to give up our mixed-ability groups in the last two years of schooling and return to some form of streaming. Such a scheme has certainly been adopted by some schools where it has been felt that the present system of public examinations militates against continuing the mixed-ability form of organization beyond the third year. We must look very carefully, therefore, at the nature of examinations to discover whether they can be compatible with a mixed-ability form of grouping or whether the ideals associated with such a form of grouping must be abandoned for ever in the fact of the stern realities of career prospects.

Internal assessment

Before we do this, it may be as well to draw a distinction between internal and external examinations and to consider briefly what our approach should be to the internal assessment of the work of pupils in mixed-ability classes throughout their school careers.

To do this we must be clear about the purposes of such assessments. Too often school terminal or yearly examinations are indistinguishable in either their form or their purpose from external examinations. Yet their purposes must surely be very different. For whereas the external examination has the largely administrative purpose of evaluating the level of achievement of individual pupils in national terms and awarding certificates that have a national currency and can be used anywhere in the country in the pupil's search for a job or a place in further or higher education, in short, of putting a final seal on the achievement of the pupil in his course, the internal examination must always be seen as ongoing, as diagnostic, as measuring the pupil's achievement at the present stage of a course that will continue next term or next year, so that its main purpose must be to guide the teacher in the decisions he must make concerning the individual's future education. In other words whereas the external examination is designed mainly to provide information for outsiders — employers and the like — the internal examination's main purpose is to provide information for the teachers themselves. In short, while the function of the external examination is largely administrative, that of the internal examination should be mainly diagnostic both of the needs of individual pupils and of the characteristics of our courses.

If this is so, there is no reason why such internal assessments should be in any way competitive and, therefore, no reason why the need to make assessments should be seen as incompatible with the principles that underlie mixed-ability grouping. School assessment should be seen as an attempt to measure the progress of individuals in relation to their own earlier achievements rather than in competition with each other. Individual pieces of work should be assessed by comparison with other work we have had from the same pupil rather than with the work of other pupils. In some cases, that there should be any work to assess may represent a marked improvement and should be credited accordingly.

This is the approach that Bruce Liddington advocates in speaking of the work of Northcliffe Community High School (1975, p. 134).

> Similarly, it is worse than useless to have a system of assessment that repeatedly places the most able children at the 'top' of the class, and the least able at the 'bottom'. This is so destructive in almost all its ramifications that I am surprised that it has lasted as long as it has in our Grammar schools and, alas, in some of our less enlightened Comprehensive and Secondary Modern schools. The only system that will work with mixed-ability groupings is one which takes into account on the one hand the *effort* a child has made, regardless of his inherent ability (thus making 'A's' not only possible but also desirable for those in the lower reaches of literacy and oracy), and also their *attainment* within the bounds of their abilities. It is only in the final stages of the fifth year that a more absolute system of attainment is necessary for external purposes..

Our concern, therefore, should be not so much to give each piece of work a mark as to give the pupil the right kind of encouragement and incentive to move on to the next stage, as we suggested when discussing the correction of the work of pupils with learning difficulties in Chapter 7, and to discover from a diagnostic viewing of the work what that next stage should be. This implies, as we have stressed throughout, playing to the strengths of pupils rather than seeking out their weaknesses. Some form of self-assessment here may be as valuable as and perhaps, from a diagnostic and motivational point of view, more valuable than the assessment made by the teacher. Progress can only really be made when the pupil himself sees the need for it and sees in detail what is involved in it. This kind of approach to internal assessment will also give the teacher greater freedom, since all pupils need not follow the same course, and will help in the development of his relationships with his pupils, since nothing can spoil such relationships more than the need to grade pupils in some kind of order of 'merit'.

At the same time, a careful record should be kept of the work of each individual so that we have an accurate account of his earlier achievements against which to assess his present work. Such an account has the advantage over the mark-book that it can include records of all aspects of the pupil's progress, his social and emotional no less than academic development. In fact, it should be a continuing record of each individual's personal achievement and endeavour. Many teachers are now finding that in this way a profile can be built up over the full period of the pupil's course at school which is of permanent value to them and of particular value to a new teacher coming to teach him for the first time.

Such information may also be of more value to a potential employer than some of the present public examination grades, since he might be more concerned with an applicant's ability to work with others in small group situations, to cope with occasional stresses or to stick at a problem until it is solved than with his knowledge of French irregular verbs or ability to handle the ablative absolute in Latin. It is interesting to reflect that suggestions were made as early as the time of the Spens Report (1938) and were reiterated by the Norwood Committee (1943) and the reconstituted Secondary Schools Examinations Council in its proposals for the establishment of the GCE that such teachers' reports should be formally built into the public examination system in order to produce a comprehensive leaving certificate, although we are still some way, it would seem, from attaining such a system. However, in time, it might be possible to incorporate a profile of this kind in the external examination system so as to get the best of both worlds, since, as we shall see, this system still lacks the sophistication to measure many of the qualities that we are coming to regard as important. There is a need for techniques that will enable us to evaluate learning over and above the acquisition of subject-matter.

We must take steps, of course, to avoid the dangers that can arise when individual teachers are asked to make this kind of subjective judgement of their pupils (Downey 1977). The present system of record-cards gives evidence of the extent to which the teacher's perception of individual pupils can be as distorted as any person's perception of another. Teachers, like anyone else, will tend to 'take to' or to 'take against' certain pupils and these initial attitudes will colour their judgement of these pupils in all contexts. Personality clashes can also occur which make any attempt at objective judgements impossible. If we are to achieve this kind of ongoing knowledge of the individual's progress, then, we shall need to employ all the techniques discussed later in this chapter, but we must remember the purposes for which we will be employing them within the school.

The public examination

If the main purpose of internal assessment is to maintain this kind of continuous supervision of the progress of individuals in order to achieve greater control over that progress and more elaborate data on which we can base decisions relating to it, the main purpose of the public examination is standardization, as we have said, and this brings us to the crux of the problem.

There is a need for the establishment of national standards for public examinations but until recently there has been a tendency to meet this need by leaving assessment to independent bodies like the universities who, through boards of examiners consisting mainly of teachers, have usually attempted to fulfil this function by setting common examination papers which, although carefully standardized, marked, and moderated, have been able to assess only relatively limited and un-sophisticated attainments and have, therefore, tended to impose a rigidity on the curriculum of the school, since teachers have naturally been tempted to limit their ambitions to those attainments that would actually be measured by the public examination.

If we are to get the kind of assessment we need for the new and more sophisticated approach to the curriculum we have described as the necessary corollary of mixed-ability groups, we need to give a good deal of detailed attention to the nature, the purposes and the principles underlying that curriculum and to the problem of evaluating our success in implementing such a programme; we need to consider some of the new techniques of assessment that have been tried in recent years to discover which are most suited to assessing our work; and we need to look at the extent to which teachers themselves should be involved in the assessment of their own pupils, since it would seem difficult to achieve the kind of sensitivity to the needs of individual schools and individual pupils that is needed without greater involvement of the teachers concerned. We must now look at these three aspects of the problem in some detail.

The key problem in any discussion of assessment is that of the relation of the methods of assessment to the curriculum model. There are very real dangers for curriculum developers here in so far as methods of assessment can govern the curriculum and, therefore, effectively inhibit curriculum development. I have seen the planning of university courses begun from a consideration of the form of examination that would be employed. That external examinations were in fact having this effect on the curriculum of the secondary school was recognized in both the Crowther (1959) and Beloe (1960) Reports in their discussion of external examinations for pupils who were not deemed able enough to take GCE ordinary level examinations. Both reports stressed the importance of freedom to experiment with the curriculum for these pupils; both asserted that for the most part teachers and schools were unable to influence the policy of the examinations these pupils were at that time taking; both realized that the examining bodies concerned were not in a position to keep in touch with research and development in teaching methods, although changes resulting from such research and development might require corresponding changes in examining techniques; and both felt that as a result external examinations were having the effect of inhibiting rather than promoting the kind of curriculum experimentation and development that was felt to be necessary and that this was to the disadvantage not only of pupils taking these examinations but also of those not felt capable of doing so.

There is no doubt that GCE ordinary level still has this kind of effect. It reflects the traditional curriculum with its emphasis on cognitive learning and its largely subject-based approach to education; it is thus based on a rather simple view of education as the acquisition of knowledge, a form of education that can be assessed more readily than most and has its attractions, therefore, both to some teachers and to examiners. As such, it provides little encouragement for curriculum development.

It is not surprising, therefore, to find that curriculum change has tended to begin at the other end of the scale. For, on the recommendation of the Beloe Report (1960), a public examination was introduced at a lower level than the GCE, the Certificate of Secondary Education, which it was hoped would be sensitive to change and would in fact encourage teachers to experiment with curriculum provision. It must be admitted that teachers were slow at first to take advantage of the opportunities the CSE offered them to experiment with examining techniques but the movement towards this is now gaining momentum.

A curriculum must be planned according to the educational needs of pupils and the educational judgement of their teachers, and techniques of assessment must be devised to fit the curriculum and adjust to it; they must not be allowed to control it. The CSE recognized this from the outset and gave teachers the opportunity to adapt the methods of assessment to their changing approaches. This they are now taking advantage of and recent developments in examining techniques can be seen as attempts to follow recent curriculum developments. It is interesting to note that some GCE Boards are now beginning to follow the lead of the CSE and are experimenting with new types of examination, sometimes after direct requests from teachers but sometimes on their own initiative. Methods of assessment, therefore, must be adapted to the curriculum. New approaches to the curriculum require new examining techniques and changes in our techniques of examining are essential if there is to be curriculum change and development.

However, the changes in the curriculum we have discussed earlier in this book and have suggested as implied in a move to a mixed-ability form of organization are highly complex and sophisticated. There is no doubt, as we have noted several times, that it is a relatively easy matter to assess the cognitive content of an individual's education, provided that we mean by this no more than the amount of information he has acquired and can recall in a given field. It has been suggested that this is one reason why some teachers find this kind of somewhat limited goal attractive.

I have claimed, however, that even the cognitive goals of our curriculum should be much more sophisticated than that. To begin with, I have stressed the desirability of encouraging individual enquiries and allowing such investigations to cross the existing boundaries between subjects. The assessment of 'integrated studies' will present us with our first problems. It was also suggested earlier that education in the full sense cannot be assessed in terms of information acquired even in an integrated way but only in terms of the development of understanding, initiation into modes of thought, the ability to operate within and between these modes of thought, to handle the concepts appropriate to them and to think independently of authority. In short, we have provided ourselves with a set of more complex cognitive goals which will require more sophisticated techniques of assessment than most of those now in operation if we are to measure our success in this sphere with any degree of accuracy.

Furthermore, I have throughout stressed the importance of creative work in education and this will lead to even greater problems of assessment. For standards of aesthetic or artistic achievement are notoriously difficult to set and, in any case, it may not be aesthetic or artistic achievement that we are after. We may be promoting it because of its therapeutic value; we have suggested it as a device for building the confidence of pupils with learning difficulties; we may be using it as a means of developing the ability to enjoy and appreciate the achievements of others more expert in these fields. All of these objectives will present us with difficult problems of assessment (Kratwohl 1964). These are aspects of the social and emotional development of our pupils which I have also argued at some length as a prime concern of the teacher and assessment in this affective sphere will be far more difficult than in the cognitive, yet perhaps more important, if not for final external assessment, certainly for the ongoing internal assessment which is of such importance to us if we are to make the right decisions with regard to the educational progress of individual pupils.

Most of these affective goals are of such a long-term nature as to make it difficult for any one teacher to evaluate his own achievements in this area with his pupils, except in the unlikely event of his having charge of them for most of their school careers. This suggests the need for a plan of evaluation at the school level, involving all of the teachers concerned with each individual. The complexities of such a plan will be apparent. We must also remember that congitive and affective goals can for the most part be distinguished only at the conceptual level. In practice, they will inevitably be interwoven. It is not possible to distinguish learning to enjoy physics from the acquisition of certain cognitive abilities connected with that subject or to speak of the development of an interest in something without assuming at the same time the acquisition of some knowledge in that area; nor, conversely, is it possible at any but an extremely superficial level to acquire knowledge in a field without any kind of motivation towards it. The complexities of these goals and the interrelationships between them make the assessment of their achievement a highly complex matter and highlight the need for continuing research in this field as well as for the employment of more sophisticated techniques of examining.

Techniques of assessment

It has become increasingly clear in recent years that the essay-type question, still a regular feature of many examination papers, has many serious drawbacks to it for most kinds of assessment and slowly it is being replaced or supplemented by other methods. It is clearly not very helpful in the assessment of the kinds of achievement we have just discussed. To begin with, it is open to inconsistencies of marking both between different members of an examining board and even within the marking of an individual examiner. There has always been some point to the hope expressed by most of us at some time that the examiner would settle down to read a particular offering of ours only after he had enjoyed a good dinner.

Furthermore, even if we can achieve standardization of marking for individual questions, we will not have ensured standardization between individual candidates. For most papers of this kind offer a range of questions from which each candidate must choose three or four. This gives a very uneven sampling of the field supposedly covered, opens the system up to the inaccurate results that arise from successful or unsuccessful question 'spotting' by candidates and, in particular, raises the problem of how we can achieve comparability of assessment between the candidate who chooses to write, say, a largely factual account of Caesar's conquest of Gaul and the one who chooses to wrestle with the complex political events leading to his crossing the Rubicon.

Such examinations do test the candidate's ability to marshal his material, to select what is relevant and to develop a coherent argument, but these abilities can perhaps be tested just as well by other methods, by timed essays or open-book examinations.

The timed essay and the open-book examination can be regarded as variations on a similar theme, since both of them allow the examinee access to research and resource material after he knows the questions the examiners are setting him. Either he is given the questions in advance and allowed to prepare beforehand to write an essay in a given time or he is allowed to bring books and other material with him into the examination. In both cases, it is assumed that the examiner will thus gain a clear impression of the candidate's ability to seek out and select information relevant to his subject and to deploy it effectively and quickly.

The individual's grasp of the factual material itself can be assessed by the use of objective tests and this is becoming an increasingly popular way of testing for knowledge of basic information. Objective questions at any level of sophistication are difficult to set but they are very easy to mark. They can be marked by clerks or even by machines, since on each question the candidate is asked to select from a number of possible answers and there can only be one right answer to each question. Thus they score very highly on marker reliability (Connaughton 1969), since there is no room for the subjective judgement of the marker, and they can cover much more of any syllabus than the essay paper, since without the need for essay writing a candidate can answer a large number of questions on any one paper. For the testing of information acquired, then, such tests have much to recommend them and they would seem to do this job better than those asking essay-type questions. They are also relatively easy to administer and this may be an important consideration if the number of pupils to be assessed rises as it is expected to. Nor should we assume too readily that such tests are only useful for the assessment of simple factual learning. This technique can be used in such a way as to measure higher level cognitive attainments too (Vernon 1964).

A completely different approach to assessment, but one which has become increasingly popular in recent years, is to base it not on the pupil's performance in tests or papers undertaken at certain set times and under certain prescribed conditions, or not solely on this kind of information, but on the work done by the pupil throughout his course.

This may take several forms. It may be based on the written work that he is required to do from time to time during his course; it may be that his teacher or teachers will be asked to make an assessment based on their view of his work over the full period of the course; or he may be asked to submit a special exercise, a project of some kind on an area he has selected to study.

This seems to offer us some hope of being able to make an assessment of the complex processes we discussed earlier and also of being able to preserve some of the individual and group assignment work and enquiry- and interest-based approaches to teaching I have emphasized. It can only be done, however, if teachers are more fully involved in the public assessment procedures, since their contribution is required either in the advice they must give to pupils on the selection of subjects for study or in the actual evaluation of their work. We must now consider, therefore, the whole question of the involvement of teachers in public examinations.

Teacher involvement

The CSE was planned from the outset in such a way as to give teachers control over it and this was done in order to prevent it from having the kind of inhibiting effect on curriculum development that we referred to earlier. Schools have been allowed to choose from three modes of examining the one they felt to be best suited to them. Under Mode 1, candidates sit examination papers on syllabuses set and published by a regional board. There are fourteen such boards in England and Wales and in each region syllabuses are designed and papers set and marked by bodies consisting mainly of local teachers. Mode 2 allows for the planning of syllabuses by individual schools or groups of schools, subject to the general approval of the regional board, but requires that pupils take external examinations on these syllabuses. A school opting for Mode 3 can set and mark its own examinations or individual assignments, subject only to moderation by the regional board. Initially, Mode 1 attracted most support, but Mode 3 has been attracting the increasing interest of teachers in recent years.

A further feature of the CSE from the beginning in all three modes has been the extent to which new techniques of assessment have been employed. Course work assessment, objective testing, project work, and oral examining have all been used in an attempt to devise more and more subtle devices for assessing the increasingly complex range of work undertaken by teachers and pupils. There are two aspects of this which are worthy of comment.

In the first place, the experience of the CSE illustrates the importance of ensuring a close link between the curriculum and the assessment procedures (Macintosh 1970). If we want to encourage teachers to experiment with the curriculum and to try new approaches, we must give them effective control of the techniques and procedures of assessment so that they can feel free to try things out without jeopardizing the chances of their pupils to attain qualifications and, as a result, putting at risk their career prospects. It is also important to make it possible to develop the curriculum and the techniques of assessments are at least as valid as those achieved by other means and teachers should have a major responsibility for assessment procedures would appear vindicated by the experiences of the CSE. To yield control of assessment to outside bodies is to yield control of the curriculum. There must be real contact between teachers and examiners, between those who plan and operate the curriculum and those who have the expertise in techniques of assessment. But if the teacher is to have the sort of freedom I have argued for to make the kinds of educational decision I have said are necessary in the individualized learning situation of the mixed-ability class, he must be confident that he can ensure that the assessment techniques will assess what he is doing and not what some outside body feels he ought to be doing.

The second aspect of this involvement of teachers in the assessment of their pupils that we must look at carefully is the implications of their responsibility in some cases for actually assessing their own pupils' work. There is some evidence to suggest (Connaughton 1969) that such assessments are at least as valid as those achieved by other means and possibly more so, since, being spread over a long period of time, such assessments avoid many of the difficulties such as illness or 'nerves' of the single occasion examination.

However, the need to make these assessments does raise a number of problems. It can make heavy demands of teachers, who are likely to be already busy enough, since it will be necessary for them, if they are to do this job properly, to attend meetings, to agree on standards, to devise appropriate tests and other forms of assessment and to carry out whatever moderation procedures are deemed necessary. It can create difficulties too in the very tender and sensitive area of teacher-pupil relationships which we have already discussed at some length. It does not require much imagination to realize how such relationships can be jeopardized when the teacher assumes the role of examiner, concerned no longer to advise and encourage but to assess and grade. For this reason, among others, it is difficult, if not impossible, for the individual teacher to achieve comple objectivity and impartiality in his assessment of his own pupils. We must also remember the need for standarization between schools and teachers in an examination of this kind, since these certificates must have a valid national currency.

Most of these difficulties can be overcome, however, if an adequate system of moderation can be devised. Such a system will protect the teacher from the final responsibility and the fear of being influenced by either a 'halo' effect or whatever its opposite is — presumably a 'forked tail' effect; it will also protect the pupil and it should safeguard teacher-pupil relationships.

There are a number of methods by which moderation of this kind can be achieved. In some cases teachers work in groups within a school or between several schools and check each other's assessments by sampling the work of each other's pupils. Sometimes visits are made to schools by moderators from regional boards to sample the work of pupils and to check the standards of assessment being used against those they have experienced in other schools. A third method is to set a common paper to candidates from all schools. The results of this test may or may not count towards the final grade, but in either case it acts as a check on the individual teacher's assessment. Combinations of these methods and other more sophisticated methods of moderation are being tried so that it should be possible to ensure that the involvement of teachers in assessment need not result in a loss of standardization of grading or of the value of such awards as national currency.

If we accept that it is possible to avoid these difficulties, then, the use of teachers' assessments in the overall grading of pupils is of enormous advantage. It provides us with a real chance of finding a method of assessing those complex goals we discussed earlier. Mode 3 of the CSE has begun to reveal techniques of assessment that hold out real hope for the future and it is not surprising to discover some GCE ordinary level boards beginning to follow the lead thus given.

Continuous assessments of various kinds, project work of the kind associated with CSE Mode 3, and many forms of test and written examination provide us, then, with a range of techniques from which we might hope to find combinations to evaluate quite sophisticated attainments in a way that will at the same time preserve national standards. Furthermore, if teachers are involved at all levels in the processes of assessment and feel that they are in control of it, this can be achieved without the imposition of rigidity on the curriculum of the secondary school and without unduly inhibiting curriculum development. The freedom of both teachers and pupils can be maintained.

A great deal of research and development is, of course, needed in this area and here again teachers need to be trained in the necessary skills if they are to play this kind of major part in the final assessment of their pupils, but this is the direction in which things seem to be going. In this context, it comes as no surprise to find proposals being put forward by a working party set up by the Schools Council for a common system of examining at 16+ (Schools Council 1971). Such a system is proposed to replace both CSE and GCE ordinary level examinations. It would hope to deal with the increased number of candidates that has resulted from the raising of the school leaving age to sixteen and with the wider range of abilities that this has also brought. It is suggested that it should be controlled by the teachers on a regional basis so as to ensure teacher control of the curriculum. It would aim to employ a number of techniques of examining, methods of assessment, and moderating procedures. Like the CSE, it would seek to grade pupils but would not use the categories of pass or fail. It would thus attempt by as many means as possible to provide pupils with a leaving certificate which did in fact give potential employers and others as much valid information as it was possible to obtain. In some areas experimental joint examinations are already in operation following the pilot schemes stated in 1971 at the request of the Schools Council.

If we have this kind of assessment, largely teacher controlled and with scope for CSE Mode 3 techniques, there is more freedom for the individual teacher and, therefore, more scope for the individual or group assignment and for interest-based and enquiry-based approaches to teaching. These methods of learning, which we have treated as an essential part of the mixed-ability class, can be maintained until the end of the fifth year without threat to national standards or jeopardy to the qualifications and career prospects of the pupils. It is still not clear, however, whether the mixed-ability class itself can survive into the final year of compulsory schooling. The proposals for a common 16+ examination envisage provision only for the percentile range 40-100, the top 60 percent of each age group, and although, as the report of the working party points out, some pupils below this range may be able to take the examination in some subjects and school-based syllabuses may cater for others, this will still leave a large number of pupils with no public examination to work towards in their final years of schooling.

Most teachers nowadays would agree that it is not necessary to start pupils on the present CSE and GCE ordinary level syllabuses in most subjects until two years, three at the most, before the examination. There is little threat to mixed-ability classes, once established, therefore, in the first three years of the secondary school. The crucial question is how far they can be continued after that, how far it is possible to prepare pupils for examinations in mixed-ability classes alongside the large number of pupils who have no such goal or incentive.

It is sometimes claimed that CSE Mode 3 can provide the answer here and can extend the freedom of schools and teachers to maintain their mixed-ability classes to the end. On the other hand, while this may have been possible in some schools when a large proportion of the less able left at fifteen, giving the others a clear run of at least one year to their external examinations, it must become more difficult now that they are all required by law to remain until the end. Many schools have already found it impossible to keep their mixed-ability classes beyond the second or third year and have, therefore, reverted to some form of streaming in the upper school. It has sometimes been argued that streaming can be 'natural' at this stage, that it can be kind of 'self-streaming', since it is based on greater evidence of the individual's abilities than has been available before and on a realistic choice by the individual of the kind of course he would like to follow with his future career in mind.

However, I have argued strongly earlier in this book the case for mixed-ability classes on the grounds of their contribution to the social and emotional development of all pupils, both the able and the less able; I have discussed the advantages of this form of grouping for the development of the kinds of relationship between teachers and pupils that education in its fullest sense seems to require; and I have stressed the importance of avoiding the behavioural problems that result from gathering all the less able pupils into a 'ghetto' class and labelling them as failures. If these arguments have any validity in the educational provision for younger children, they must have at least equal validity when we come to consider that of older pupils. Indeed, we know that the older the pupils the more difficult they are to handle in this kind of situation. We should be reluctant, therefore, to break up the social groupings that have existed earlier in the school and to throw away the advantages we can hope to have gained from two or three years of mixed-ability groupings. We must finally turn, then, to a consideration of how these groupings might be maintained in the final years of compulsory schooling without detriment to the examination and career prospects of any pupil.

The mixed-ability class in the final years

In looking at this question, it might be as well to begin from a consideration of some of the things those pupils whose general ability does not seem to warrant their taking external examinations should be doing in the last years at school. As a general principle, they should be engaged in activities that they can themselves see as having value for them. If they are not to become a threat to the good order of the school, they will need to be kept occupied and, for all the reasons we gave in Chapter 8, they can only be kept properly occupied if they can be persuaded to accept the value of what they are doing or can be allowed to do the things they regard as valuable. This is one good reason for continuing to allow them to work through their interests and on their own individual or group enquires. We shall not get far if we take the attitude of the teacher I once heard complaining, 'I can't teach these fourth years any physics. All they are interested in is motor-bikes.' A lot can still be done to promote their education through the interests they reveal and this should continue.

In addition to the continuation of this kind of provision, however, there are certain specific things that they all become interested in during the years preceding their final departure from school. In particular, they become interested in their future work and all that is associated with it. There is no doubting, therefore, the need for their curriculum to contain a large vocational component at this stage. There are at least three elements in this provision. They need to be provided, where this is possible, with the beginnings of some of the specific skills they may need in the job they will enter; they need a more general introduction to the world of work, to the basic features of working life; and they need extensive guidance in their choice of a career.

In most situations it may not be possible for the school to do a lot to provide pupils with specific industrial skills. In many cases they may be going into too many quite different careers for this even to be considered. Where there is a dominant local industry, however, to which most of them will go, it would seem only good sense to give them whatever help the school can give in preparation for this. Obviously, the workshop staff have a major contribution to make here and it is always noticeable how seldom disciplinary problems arise when boys of this age are engaged in the kinds of activity that modern craft teachers can involve them in. With girls the domestic science staff will be similarly involved and, it is hoped, to the same effect. We should be wary nowadays, however, of being too ready to allocate sex roles in this way. Unisex is as much a feature of education as of society as a whole and schools are producing as many male cooks as expert craftswomen. Schools should do whatever they can, in the light of local conditions, to help pupils in this way.

Linked courses with colleges of further education can be particularly valuable here since they can enable pupils to get the kind of vocational course most schools are not equipped to provide, they can bring to the pupil's attention the range of postschool education that is available and they can enable him to spend some of this time in a situation where he can have more personal responsibility.

Much more important, however, is the provision of courses of preparation for the world of work. This is something all pupils need and if the school does not provide them with it no one will. It is a major step to move from the relatively cloistered atmosphere of even a large modern comprehensive school to a factory, a shop, an office or a typing pool and this is an experience few teachers have had, since transition from school to work for most of them was cushioned by a period of time at college or university. To begin with, one is in school as of right; one is at work only if one can show that one has a contribution to make. The school is, for the most part, a society of young people; at work there will be people of all ages, from sixteen to sixty-five. There is often an abrupt change of values, of attitudes and of atmosphere. On top of that there are things like income tax, insurance contributions, benefits, pension schemes, unions and a hundred other new things to cope with. This is a transition that cannot be made smoothly unless the school helps pupils to make it.

The means that they must be provided with a lot of basic information about conditions of employment and the like and must be prepared in every way possible for the new situation they are about to enter. Some schools have experimented with the simulation of work conditions in the school. They have required the older pupils to 'clock' in and out and have introduced production or bonus-incentive schemes into their work. The 'spin-off' advantages of this kind of approach are not to be scorned too readily by teachers concerned to motivate pupils in their final years at school.

Another way of preparing pupils for work is to arrange for them to visit places of work in the locality. Such visits should, if possible, not be merely guided tours conducted by the manager; there should be an opportunity for them to mix with and talk to the men and women on the shop floor and to find out at first hand what it is really like to work there. If it can be arranged for them to visit such places on a more regular basis and perhaps actually work in them for part of the week, this will be even more advantageous, since a phased transition from school to work then becomes possible with school support and super-vision of the intermediate phase. Such a course may be extended to or developed from a wider course of education for citizenship — certainly there is much to be said for attempting to put all of this into a broader context — but the important thing is that the school should accept and face up to its responsibility in this area.

As a part of such a course and as an integral part of the provision made by a school for its pupils, guidance in the matter of choosing a suitable career is of absolutely vital importance. Stories of boys and girls going through a dozen or more jobs in the first months after leaving school are not uncommon nor are they particularly surprising when one considers the inadequacy of the provision made for advice on these matters by many schools. This must be a major concern of the pupil in his final year, if not before, and we cannot really expect to win his respect or his attention if we do not take it very seriously.

To take it seriously involves a lot. It is important, first of all, to provide information about the jobs that are available, the career prospects that they offer, the conditions of work and the qualifications needed for them. Such information can be provided by pamphlets, by visits from representatives of the firms and sometimes by films and other publicity material of this kind. All of these facilities should be made available to the greatest extent possible.

To provide all of this is only to scratch the surface of the problem, however, since none of it will help the pupil to discover what it is like to work in this factory or that office nor to know whether he is suited to that particular kind of work. Provision of information needs to be supplemented by the kind of visit to local places of employment we have just described and, if possible, opportunities to try working in them to get the feel of them.

None of this will be of full value, however, if the pupil is not given all of the help he can be given with the task of discovering the kind of job he is personally suited for. It is not enough to ask a pupil in his last year or his last term what he wants to do or to push him towards that which he is qualified to do. Dissatisfaction arises when a young person finds himself in a job that may have appeared attractive but which he quickly finds is not his thing. If pupils are to be helped with this and given appropriate advice, then teachers need a great deal of data about them as individuals. This is another way in which the profile can be of great value, but it must be a profile drawn up with this as one of its specific concerns, since it must include the results of aptitude and interest tests, not necessarily highly sophisticated, that have been used over a period of time to build up a picture of each pupil's aptitudes, interests, and preferences as well as his capabilities. Without this sort of data it is sheer arrogance to assume that careers advice can be given, but such advice must be given and it must be readily available.

Nor can this be left until the last few weeks or the last term before the pupil leaves school, when the youth employment officer calls to give each pupil an interview that the magnitude of his task makes necessarily brief. Advice and information should be available and should be given throughout the last two years of schooling at least. Some would wish to begin much earlier than this and start children thinking about these issues almost from the outset of their secondary-school careers. Certainly, the matter of building up an aptitude and interest profile cannot be begun too soon and it is most desirable that formal arrangements be made for careers 'lessons' throughout the secondary-school course.

If this is to be done properly, again it will place a heavy burden on teachers. In this field, however, it is relatively easy to delegate prime responsibility to a specialist careers master or mistress. This does not mean that the other teachers can forget their responsibilities here, since, like moral education, vocational preparation in its broadest sense must be seen as a responsibility shared by all teachers in the secondary school, but it does mean that much of the hard work collecting information, testing for aptitudes, arranging visits and so on can be done by one man or woman with a timetable and other provision to enable him or her to do it thoroughly. It also means that all pupils can benefit from the advice of a member of staff who himself has had industrial experience, since first-hand experience of this kind would seem to be an essential qualification for such a post.

The whole area of vocational preparation, then, is a huge one and one that must loom very large in the final years of compulsory schooling.

A second area of great importance is that of community service (Schools Council 1968). I have already referred to this in stressing the need of pupils with learning difficulties in particular to feel wanted and needed and to be shown ways in which they can make a valuable contribution to their community. Many schools have found this a valuable and important element in the education of their fifteen- and sixteen-year-olds. It can give them a sense of purpose that will obviate some of the behavioural problems that can arise at this age; it contributes also to their social education in that through it they can learn to care for others, and to put themselves into the position of others, so that they learn something about the interdependence of men; it can also be developed into a course of some academic substance since active involvement in community service can act as a basis for a social studies course that may be more worthwhile than many such courses that are being offered at present.

Indeed, this interest in community service must be seen as one aspect of the interest that is shown by all pupils at this stage of their development in the world outside their own immediate environment of school and home. It is at this time that their attention turns outwards and they become interested in social issues, in national and world problems, in political and moral questions. It is at this time too that they are most in need of moral guidance, since developing sexual maturity and the approach of adult status bring feelings of independence and personal moral problems for which personal solutions must be found. They can no longer be buttressed from such issues by their parents; they must face them themselves. The guidance they need, therefore, cannot take the form of moral precepts; they must be given opportunities for and help with informed discussion of these issues if they are to be able to reach sound and considered conclusions on them, at least on those that affect their own lives and behaviour.

In short, this is the time when their moral education should be reaching its climax and each pupil should be being helped towards the greatest possible measure of moral autonomy. Thus, it is the aim of projects such as the Schools Council's Humanities Curriculum Project, to which we have already referred several times, to help teachers to give their pupils this kind of opportunity and to provide them with some of the resources needed if they are to do so. Other material is also available to teachers who wish to help their pupils with this kind of exploration. Again, it is an area which, because of its importance to the pupils, schools cannot afford to ignore.

Furthermore, it is an aspect of the development of their pupils that all teachers must recognize as a shared responsibility. For moral education, like language learning, goes on in one form or another all the time, so that no teacher can afford to ignore his own contribution to it. Indeed, there is a good deal to be said for seeing all three of these elements we have just listed separately as a unity, as being merely different aspects of the overall personal development of the pupil and as different dimensions of the pastoral responsibility of the school for all of its pupils. If this is so, then there is a strong case for organizing it on a school-wide basis and involving every member of staff in it in one capacity or another. Only in this way can we be sure of creating a climate of pastoral care that will be reflected in the organization of the school as a whole and in the approach of every teacher in it rather than developing a system that will perhaps be of limited value if left to one or two 'specialists', whether teachers or counsellors, and not incorporated in the total structure (Schofield 1977).

All of these elements in the work of the final years of schooling can be developed into viable courses and even into courses leading to publicly approved qualifications. They should not be seen as necessarily something that is done to 'keep them off the streets' or to provide them only with the very practical help they need. All are very important elements in the education of these young people and can provide a basis for worthwhile courses leading, particularly via Mode 3 techniques, to recognizable qualifications for pupils who might otherwise have been regarded as 'unexaminable'. It is not necessary to see the public examination as concerned only with traditional academic subjects. Anything we regard as being educationally important we should be prepared to recognize and to offer pupils qualifications for if they reach a level of attainment that warrants it. If we take this view, it becomes possible to find areas in which public qualifications are within reach of a great many more of our pupils than appear able to achieve them at present and thus to provide them with a valuable source of motivation for their efforts during their final year.

However, in listing the provision that needs to be made for less able pupils, I have touched on nothing that is of any less value to those who are to take public examinations and even to stay on for advanced courses. He would be a brave man who would claim that such pupils do not need opportunities to discuss moral issues, the social education that comes from community service, or the confidence and security that can come from proper vocational guidance. Indeed, it is often claimed that university entrants and university graduates show the lack of proper vocational guidance more than most, since they have so often not been advised on the subjects to study or the courses to take with particular careers in view. It may be that in the fourth and fifth years these pupils need a differently balanced diet from that of their fellows who will leave at sixteen, but it is hard to say that they need a totally different kind of provision. Furthermore, we must avoid those dangers we discussed at some length in Chapter 2 of allowing the generation of totally different curricula even at this stage.

For these aspects of their work then, at least, there would seem to be a good deal of merit in retaining the mixed-ability classes we will have established in the early years. Such classes will also provide quite suitable groupings for a number of other areas of the curriculum — games, P.E., art and possibly others. We would be well advised to avoid as far as we can the generation of totally different curricula for different 'grades' of pupil with all the attendant problems of the stratification of knowledge (Young 1971) and behaviour difficulties (Hargreaves 1967, 1972; Lacey 1970) that such a policy will engender.

How far can these groupings be maintained at this stage for other areas of work, especially the more academic areas? This is the crux of the problem facing the schools and a number of different solutions have been explored.

To begin with, an increasing number of schools are finding it possible to devise courses which lead to CSE qualifications — with parallel GCE ordinary level courses in some cases too — but which, being based on individual assignment and enquiry methods, provide scope for a wider range of children than would normally be expected to reach this kind of level. Secondly, where parallel CSE and GCE courses have been made available, candidates have been entered for one or the other examination according to their level of achievement. Humanities courses and other courses of 'integrated studies' seem to offer particular opportunities here. The next stage of this process is the designing of syllabuses to be submitted to both CSE and GCE boards.

Such a syllabus has, for example, been devised and submitted by the staff of Northcliffe Community High School (Steels 1975). Described in the submission as 'not a watered down GCE syllabus for submission to the CSE Board, but a genuine syllabus which puts the future educational needs of the children before every other consideration', it is a course of integrated studies on a series of topics influenced by and based loosely on the Schools Council's Humanities Curriculum Project, and so planned as to encourage a degree of individual pupil choice within a broad framework of common ground and to allow grades to be separately determined for the four subjects principally involved in the project, English, geography, history and religious education. Evaluation is based on the continuous assessment of course work and a final dissertation.

Such a scheme points conclusively to the need for a common system of examining at 16+; it also draws attention to opportunities that the mixed-ability class can provide for teachers who are prepared to approach the education of their less able pupils with imagination and confidence. The 'self-fulfilling prophecy' can work to raise the level of pupils' achievement as well as to lower it and pupils of all abilities can continue to work together throughout their school careers.

However, there may well still be areas of the curriculum for which for a variety of reasons this kind of solution will not work and a differentiated provision will need to be made for different groups of pupils. Most schools have in fact discovered that for some subjects at least a system of 'setting' has to be introduced at some stage in the upper school. While the introduction of such a system must obviously raise fears of those social difficulties we have referred to so often, there are a number of ways in which it should be possible to avoid this.

In the first place, if we have retained our mixed groups for many other purposes, such as those we have just been discussing, the effects of any 'segregation' for specific purposes must be minimized. In fact, this will be no more than another version of that system of withdrawals we discussed earlier. Secondly, many of the divisions that we may have to make at this stage in the career of our pupils will be based on their own choices anyway. To a large extent, therefore, the allocation of pupils to special classes at this level will be a result of their own choices based on their interests and ambitions rather than merely on ability. The question of ability must, of course, enter into such choices but we have now reached a stage at which most pupils have a realistic view of their abilities and will often make sensible choices of their own. Clearly, it will make a lot of difference if a pupil, albeit after advice from his teachers, chooses a course or opts out of a course on the basis of his own appraisal of his capabilities rather than that of someone else. Thirdly, an important difference between setting and streaming is that the former is based on pupils' abilities in specific subjects and the grouping is used only for each of these subjects, so that there is no suggestion in it of total lack of ability or rejection. If all of these factors are borne in mind, there is no reason why any system of setting that is introduced should be accompanied by the return of social disadvantages or behavioural problems.

Indeed, there may well even be some advantages to be gained from this kind of arrangement. In some subjects, for example, it has made possible the development of different kinds of examination programme to suit different groups of pupils even within the same subject, not only separate GCE and CSE courses but also courses with a completely different kinds of emphasis — within French, for example, French language and European studies courses (Walmsley 1975). This variety of offering clearly has great advantages but would be very difficult to implement within the same class or group.

A further point that is worth noting, however, is that setting seems to work more effectively where it is built on a mixed-ability foundation. One of the advantages that such a foundation seems to provide, for example, is a wider net to catch a broader group of able pupils. If we can take French again as an example, it does appear that many able and interested pupils are gathered in by a mixed-ability organization in the lower school and thus attain the 'top set' in the upper school (Walmsley 1975). These are the pupils who in a streamed system would have been placed in a lower stream and either might never have been given the opportunity to learn French or might have been offered a watered down syllabus and thus suffered the disadvantages of low teacher expectation we referred to earlier. The same phenomenon was observed and recorded by Arthur Young (1975) in relation to those pupils who achieved success in the combined GCE and CSE humanities course at Northcliffe Community High School.

Setting, then, can be a very useful device in a number of different ways and certainly would not seem to be incompatible with the basic ideals of the mixed-ability class. It is another example of that flexibility of grouping I argued in Chapter 1 as the major rationale for this system. The important thing is that we should use it sensibly and sensitively in those areas where it seems necessary, while maintaining as an overall aim the retention of mixed-ability groups for as many purposes as possible. The mixed-ability class thus becomes a firm and stable base upon which many other kinds of grouping can be built without detriment or disadvantage to the basic ethos we wish to establish and maintain.

Unless we can retain the mixed-ability class to the end, we will create classes of leavers that will be a constant source of disruption to the life of the school and the extra year will become the hazard to life and limb that many teachers expected it to be. Nor will we have attained anything of value if our mixed-ability groupings are abandoned after three or four years. For with them we will have abandoned our values and our ideals and the pupils will see them for the empty things they are. We need to be consistent in our attitudes; we need to see secondary education, and indeed all education, as a continuous process with the same basic principles throughout. Just as we must not allow external assessment to determine our educational goals for us, so we must not allow it to govern the administrative procedures we adopt to achieve them, since, as we have seen, these procedures themselves have an educative function and, if our purposes are to be achieved to any extent, the whole educational environment must be directed towards their achievement.

Summary and conclusions

Acknowledging the need for regular appraisal of pupils' progress and especially for examinations of a national validity, we have in this chapter looked briefly at internal assessment procedures, suggesting that the major characteristic of these should be a concern with a progress of the individual rather than the grading of pupils in relation to and in competition with each other. We then considered some of the major aspects of the public examination system, noting the danger it threatens of restricting the freedom of both teachers and pupils and inhibiting the kind of curriculum development I have claimed elsewhere must be seen as a prime feature of mixed-ability grouping. We next discussed some of the examining techniques that are available to us and in particular some ways in which teachers themselves can be more closely involved in the assessment of their own pupils, suggesting that this kind of teacher involvement along with a more sophisticated range of techniques can be used to offset some of the inhibiting effects of the public examination.

Finally, attention was turned to the difficult question of how it might be possible to retain mixed-ability classes to the end of the period of compulsory secondary education in order not to lose those advantages it has been claimed such classes can bring. It was suggested here that the dangers of generating two completely different curricula at this stage as well as of creating different social groupings were to be avoided and that this could best be done by identifying those areas of common interest, concern and value to all pupils and offering these to them all in the same mixed-ability classes. We noted too that there would be many other areas of the curriculum that they could continue to pursue in these same classes. Lastly, it was suggested that where some grouping by ability was deemed to be necessary, such divisions could best be made in relation to specific subject areas by a system of setting, since such a system avoids most if not all of the ills of streaming and may even have some positive advantages.

Again, therefore, we note the essential need to be flexible in our approach and in our organization and to establish this kind of flexible system on a continuing mixed-ability base.

CHAPTER 10

A FLEXIBLE SYSTEM

Flexibility was the theme on which we ended our discussion of the arrangements that can be made for all pupils in the final years of compulsory schooling in the last chapter. With it, therefore, we came full circle to our point of departure. For it was flexibility that was put forward as the main feature of the rationale for mixed-ability grouping that I tried to establish in Chapter 1. In fact, flexibility has been the theme throughout, the key to all the issues and problems we have looked at in the intervening chapters — organizing individual and group assignments, grouping within the class, dealing with those pupils who are experiencing learning difficulties, arranging for suitable public examination procedures and so on. It is right and proper, therefore, that this long discussion of teaching mixed-ability classes should end with some more extended reference to it.

There are two things that need to be done to round off this discussion with an examination of what is implied by arguing for flexibility as the central feature of mixed-ability grouping. The first of these is to consider briefly one or two areas which have not hitherto been touched upon where it has particular advantages to offer. The second is to look at some of the implications it has for the continuing development of education. For this last point was one that I stressed in concluding my attempted rationale at the end of Chapter 1 and it is one to which we must return in spiral fashion now that we have extended our understanding by an examination of many of the other dimensions of mixed-ability teaching.

Initial allocation to classes

Little has been said so far about the basis upon which mixed-ability classes might be created in the first instance, but here again a flexible approach will prove of advantage. We have noted the different criteria that might be used and are in practice used by schools in making their initial allocation of pupils to classes. Some adopt a totally random procedure, basing the distribution on alphabetical order or some other such neutral criterion; some use dates of birth; others attempt to make their selection in terms of friendship or neighbourhood groupings based on primary schools of origin; yet others aim for a deliberate mixing of abilities by using IQ scores and the results of attainment tests or primary teachers' assessments in an attempt to create classes which are of mixed ability in a very literal sense.

This variety is reflected in the following two accounts of how this allocation has been made in two girls' schools. First of all, Margaret Horne describes the procedures used at Fairlop Secondary Girls' School (1975, p. 43).

> Each form group is composed, on entry, of pupils of every level of ability from the brightest to the dullest. We arrive at these form groups by information from Primary school record cards and by visits to the Primary school in June and July by our first year tutor and as many teachers as we can spare. At the same time as we organise a spread of ability for each group, we also arrange for every girl to be with at least one friend. We draw from about ten contributory schools and there will automatically be many 'strangers' in the newly composed forms. It seems helpful and reassuring that familiar faces shall also be found.

A rather different approach is described by Elizabeth Hoyles as emerging from the experiences of Vauxhall Manor School (1975, p. 50).

> As time went by, the problems of grouping became more difficult. Originally it had been decided to group the children according to their date of birth so that it was possible to find in one class children of September/October birth dates and in another class children of

July/August birth dates. These last children, who always face difficulties because of their shortened Primary School experience were still at a disadvantage when they came into the Secondary school and, as time went by, began to be considered as 'the less able group'. In addition, the children in the classes that contained the September/October/November birth dates were under serious pressure from their peers when reaching statutory leaving age and there were always considerable problems for those who were able to stay on but who felt themselves required by their group to leave school and to start work.

The whole question of grouping, therefore, came into reconsideration because of this situation. Eventually, it was decided that perhaps a fairer way might be to arrange classes according to the initial letter of the child's surname, arranging the entrants in alphabetical order and dividing them up into eight classes. This system had the advantage of showing the children that they had been put into classes for reasons that had nothing to do with their own ability, whether based on intelligence tests or on academic achievement in the Junior school. It is also the case that in an eight-form entry school such an arrangement is statistically likely to provide equal classes of widely mixed ability.

In both cases, it is worth noting the degree of flexibility employed. Either more than one criterion of choice is used, as in the combination of attainment and friendship patterns at Fairlop School, or the system is quickly changed if it proves to have flaws that were unforeseen, as was the case at Vauxhall Manor School. Furthermore, these groupings once created are not regarded as fixed for all time. For Elizabeth Hoyles goes on to tell us (1975, pp. 50-51):

This does not mean, however, that all the classes are of equal social mix, nor does it mean that every class is exactly the same, because without fail, year after year, there have been classes that have been more difficult to deal with than others, classes that have been more fragmented, and on the other hand classes that have more quickly settled down into a composite whole.

It seems likely, therefore, that there is always a necessity to find some way, perhaps during the course of the first half-year, of doing some regrouping in order to spread the opportunities available to all pupils over the full range, since it is possible, if classes that are difficult remain together, that they will, in fact, benefit very little from all that is offered them in the school.

It is important, of course, to provide a secure base for pupils, especially at this stage in their schooling, so that we must beware of overdoing the regrouping, but to allow a measure of flexibility here, both to correct things that may have gone wrong and to adjust and modify our procedures to try to avoid a repetition of them, is clearly very important too.

These same considerations must be kept in mind when looking at the question of regrouping generally. We must try to avoid, especially with younger pupils, a situation where they are members of so many different groups that they 'belong' to none. On the other hand, we have on several occasions had cause to note the advantages that such regrouping offers in dealing with the problems of the poor reader, for example, and pupils with other kinds of learning difficulty, in catering for the highly gifted pupil who may need to be extended in certain fields in such a way that may not be possible in the mixed-ability class and, as we saw in the last chapter, in making suitable arrangements for the preparation of pupils for public examinations.

Particular subject areas

Another advantage that this kind of flexibility of grouping between classes offers which we have not so far considered is the ease with which it is possible to make different arrangements for different subject areas. We noted in Chapter 3 that some teachers in certain subject areas, notably modern languages, mathematics and science, have taken the view that the nature of their subjects makes the teaching of them to mixed-ability classes impossible. It was suggested then that this is not due to anything inherent in these subjects, since there are other teachers of these same subjects who do not share this view, and that it might be necessary for teachers to think again about the point and purpose of the study of these areas of the curriculum before taking too dogmatic a line on this issue. However, it is obviously true that if teachers believe their subject cannot be taught to mixed-ability classes, then they themselves will be incapable of teaching it successfully to such classes, so that some other arrangement has to be made if we are to avoid creating a situation in which some teachers are forced to adopt a scheme they do not believe in and some pupils, as a result, receive poor service in a major curriculum subject.

Again the flexibility offered by the kind of mixed-ability system that is being advocated provides the solution. For there is nothing to prevent the introduction of setting for these subjects even in the first year of the secondary school, if some subject departments want it. Such an approach has been adopted by many headteachers to deal with departments whose members were opposed to the introduction of mixed-ability classes or who merely wished to hold back until they could see how other departments made out. Obviously, this is a sensible way of dealing with this problem and quite the best way of giving the reluctant ones a chance to think things over. It is not necessary for every subject to operate the same scheme nor for the same class groupings to be used for all purposes.

These points are illustrated very clearly in Andrew Hunt's account of the arrangements at Sir Leo Schultz High School (1975, p. 68).

> This team-teaching situation in the school could have led to the emergence of groups other than mixed ability — setting, for example, was possible theoretically (although it could only have been over one quarter of the school population in any year). At that stage in the school's development, I believe that this was the right thing to do. Rather than foist mixed ability teaching willy-nilly on uncertain and unready colleagues. I preferred to build on the mixed ability work which had been going on from the start in most of the Humanities subjects, as well as in Arts, Crafts and Technical Studies. The point is that mixed ability teaching became so much the norm in the school that staff generally were loathe to depart from it, seeing in it so many of the answers to problems of selection inside the school (with the concurrent rejection of some pupils). Thus although Languages notably, and Mathematics and — to a lesser extent — Science, all flirted with more traditional forms of grouping pupils, by 1969/1970 all work in the intake year was based on mixed ability inside the house population.

Many people would want to claim that the need for regrouping in the early years of secondary education can and should be kept to a minimum, partly because of a conviction that it is not really necessary for any subject at this stage to endeavour to achieve homogeneous groups and partly because of the need to provide a reasonably stable unit that each pupil can come to belong to, feel a part of and thus gain security from. It is also at this level that the problems of selection to which I referred in Chapter 1 are most significant and troublesome. Furthermore, in Chapter 9, attention was drawn to some of the advantages that may accrue to a later system of setting if it is built on a fully mixed-ability base. However, if the teachers of certain subjects remain convinced and do not accept these arguments as valid in relation to their own area of responsiility, although we may, and perhaps should, try to persuade them to rethink their attitudes and position, we must respect their views and make due allowance for them. A proper mixed-ability organization should be flexible enough to provide, at least in part, the alternative scheme or schemes they want.

The transition to mixed-ability classes

It is this facility that offers also the possibility of a smooth introduction of mixed-ability classes. The change from a streamed to a totally unstreamed system is a major upheaval for any school and can be a traumatic experience for head, staff and pupils, not to mention parents, governors and others who are interested in the school's welfare. The manner in which the transition is effected, therefore, and in particular the kind of advance planning that is done are crucial.

There are a number of ways in which a change such as the introduction of mixed-ability groupings can be brought about in a school. At one extreme, it can be imposed by the headteacher in an autocratic manner regardless of the feelings, beliefs or opinions of other members of staff. In a few cases, it is clear that this method or something approaching it has been used, but it is also clear that it is never successful, since, as we have seen, teachers cannot make a success of a system they do not themselves believe in and some will in any case indulge in acts of deliberate sabotage. At the other extreme, it can be brought about by pressure from the teachers themselves or a body of them who succeed in convincing the head of the need for such a change. Again this has happened in a number of cases and, provided that the head is genuinely convinced and prepared to support the change, a start like this obviously promises more hope of success. In most cases, the reality lies somewhere between these extremes. What is important, however, is that, regardless of who takes the initiative, any such innovation must begin by an attempt to convince all or most of the people affected by it of its value. In seven of the nine schools that were studied in the recent ILEA Inspectorate's survey, to which I referred in Chapter 1, for example, the process began with 'long, continual and thorough discussion by a high proportion of the staff during the period in which decisions were taken' (1976, p. 16).

However, the teachers are not the only people who need to be persuaded of the value of such a change and, indeed, one reason why they must be persuaded is in order that they may help in the process of convincing others, such as governors and, above all, parents. The experience of Margaret Horne at Fairlop School is salutary here (1975, pp. 43-44).

> For our pupils these two changes into smaller and unstreamed groups seemed pure advantage but we had not reckoned with the wrath of a few parents whose children would have been 'A' but who now were unlabelled. They sorely missed the gratification that they were denied by our so-hopeful new arrangements and they raised long and vigorous

protests that would have daunted all but a very committed team of teachers. Misgivings and forebodings were uttered and, had not the pupils in the main been happy to come to school, we should have lost many girls to the more formal neighbouring schools. Five years later improved examination results told their own story and our methods were proven effective, if not yet popular with our parents.

Parents and governors are demanding, with some justification, at least in the case of the former, an increased say in the way in which a school organizes itself and plans its curriculum, so that it is right and proper, as well as sensible and politic, to discuss any major change of this kind with them and to endeavour to win their confidence and support for it. This kind of confidence and support can best be won by putting parents fully into the picture, by telling them what one is hoping to achieve by the change and even by involving them individually or, perhaps on occasion collectively, in some of the work.

Lastly, it has been stressed continually that teachers need not only that conviction of the value of mixed-ability teaching that will both come from and lead to a deepened understanding of its implications and of its educational significance; they need also to develop a wider range of teaching skills. The ILEA Inspectorate's report has put this very clearly (1976, p. 17).

> There appears to be general agreement that mixed ability teaching makes demands on techniques, methods, materials and standards markedly different from those operating in the schools in which teachers themselves were educated; different from those usually acquired in Colleges of Education and University Departments, at least until very recently; and different from those practised in schools where streaming or banding is the order of the day.

It has, of course, been a major aim of this book to assist teachers to develop these new techniques. There is no doubt that this kind of change has been far smoother and more effective where formal opportunities have been made available to teachers to acquire them through adequate in-service provision. This again emphasizes the need for careful and thorough preplanning and suggests that a good deal of preliminary work of many kinds is necessary. All of this preparatory work will contribute towards a smooth and effective transition.

If, however, in the face of all of these major tasks, the transition is to be as painless as possible, if it is to be, in Arthur Young's words (1975, p. 29), 'evolutionary rather than revolutionary', there are enormous advantages in introducing it stage by stage in a carefully phased and gradual manner.

Again, therefore, it is the potential flexibility of mixed-ability grouping which offers a range of options here and makes possible the introduction of such a scheme step by step rather than at a stroke. There are at least three ways in which it does this.

In the first place, as most teachers have realized, it makes good sense to introduce it initially for first-year pupils only. In this way older pupils are not expected to learn to work in a new and different way in the same school and only one section of the school and a small proportion of teachers are involved at the outset in a new system. Furthermore, disadvantages have been experienced in those cases where attempts have been made to introduce it for pupils who have already settled into a different pattern of working. There is a lot to be said for developing this kind of individual approach to study over the years; certainly, it appears to be very difficult suddenly to introduce it when pupils have got well used to another kind of approach. Introducing such a change in this way allows the new system to grow through the school with both pupils and teachers gradually acquiring the experience and the ability to work in the new ways it requires.

Secondly, it is possible to introduce mixed-ability teaching at first in a modified form by the device of 'broad banding'. This is a system in which pupils are grouped according to their measured ability into several (usually three) broad bands and 'mixed-ability' classes are created within but not between these bands. Such a scheme ensures that teachers are not faced on their first experience of mixed-ability teaching with a class containing the complete spectrum of abilities, so that they can thus learn to cope with a spread of abilities with classes in which that spread is not too wide.

Thirdly, it is possible to resolve the problem of particular departments or teachers in particular subject areas who, as we have just seen, remain unconvinced that their subjects can be taught adequately to mixed-ability classes. In almost all schools there have been certain subjects for which some kind of grouping by ability has been retained for this sort of reason or, at least, this kind of facility offered to some subjects. In any case there would again seem to be advantages in phasing the introduction of mixed-ability classes by establishing it first of all in some subjects only and especially in those where grouping by ability seems largely unnecessary and perhaps even superfluous.

All of these arrangements should be seen as interim steps towards the introduction of mixed-ability classes in the full sense; all of them offer, singly or in combination with each other, a range of much needed opportunities for a gradual phasing of this change; and all of them are aspects of that flexibility that is, or should be, the central feature of any system of mixed-ability grouping that is introduced.

It has been a major theme of this book that that kind of flexibility is necessary to meet successfully all the multifarious goals of the school, to allow for the provision of an education tailored to the particular requirements of each individual pupil, an essential element in any truly educational process, and to make possible continuous and smooth curriculum change and development. It has also been a theme that it is mixed-ability grouping which, if properly conceived and executed, offers the greatest degree of flexibility and thus the best framework for continuous adaptation to changing social demands and needs, for proper curriculum development and, as a result, for education itself.

The teachers

If education is to move with the times, this kind of flexibility is essential. However, a system such as this must place greater responsibility and power in the hands of those whose job it is to operate it. This must be recognized as an inevitable feature of those changes in both schools and society itself which we discussed in Chapter 2; it is part of that shift Basil Bernstein has described in considering some of the implications of the move towards the open school and the open society (1967). It offers less scope rather than more for external control and direction and gives teachers more responsibility both for the education of individual pupils and for the continuing development of educational provision as a whole. There are two aspects of this which we must briefly note in concluding our discussion of some of the practical and theoretical implications of the move to mixed-ability teaching.

In the first place, this may appear to run counter to the current trend towards making teachers more accountable for their practice. As we saw in Chapter 1, there has grown during recent years a feeling that teachers should not be allowed too much freedom in deciding either the matter or the manner of education and this has been, not surprisingly, a major factor in that reaction against mixed-ability grouping we also noted there. There is no denying that what is being advocated here as the only effective way of operating a mixed-ability scheme must have the effect of increasing rather than limiting the freedom of the teacher in relation both to the education of individual pupils and to the continuing development of education in general.

However, there would seem to be no other way of ensuring that educational provision can be tailored to the individual pupil and that there can be continuous curriculum change and development in a manner that is free of the worst kinds of friction (Kelly 1977). We have had cause to note frequently throughout this book how crucial the individual teacher is to the education of all pupils and, whether we like it or not, we have to allow him the freedom to fulfil that responsibility according to his own professional judgement, since there is no other way in which pupils can be offered anything fully worthy of the name of education. This is why those demands for greater accountability that I have referred to, when spelled out in more detail, always emerge in utilitarian, economic or vocational terms, since this is the only kind of provision that schools and teachers can make on the dictation of outside bodies.

It does not follow, of course, that in the kind of structure that has been described and, indeed, advocated teachers are to enjoy complete licence. An acceptance of this kind of approach to education, which, it has been argued, is integral to a mixed-ability form of organization, is not incompatible with the notion that many decisions concerning the balance of the curriculum and its overall content can be made in advance and with a full recognition of the views of those people outside the teaching profession who might be felt to have a right to comment on such matters.

On the other hand, in this kind of teaching context, a good many decisions must be made by the teacher on the spot. And clearly, the decisions that teachers make in this way cannot be approved in advance, the teachers themselves cannot be operated by remote control and the educational needs of individual pupils cannot be adequately catered for from outside the school.

Again, however, it should not be assumed that, even in respect of decisions of this kind, teachers are not to be accountable. It is only if we define accountability in terms of prior approval for all decisions made that this is so. They can and should be required to justify such decisions and actions, but the etymology of the term 'accountability' would seem to imply that such justification may be *post eventum*. We must trust all teachers, and particularly those whom we have appointed as heads of schools or departments, to be able to exercise their professional judgement wisely and sensibly in the best interests of both our children and society; and we must not allow the occasional failure to blind us to the frequent successes. Unless we can do this, our schools will become institutions merely for the training and socializing of our young and little worthy of being called education will be possible.

The second major point that must be made is one that follows closely on this claim. For if teachers are to be given this kind of responsibility and demands made of them such as have been described, it is clearly in everyone's interests that they have every opportunity to prepare themselves as fully as possible for their tasks and to continue to improve their skills, their knowledge, and their understanding throughout their careers. For in the last analysis any education system will stand or fall by the quality of its teachers and the system we are advocating will demand teachers of a very high quality.

Initial training courses, then, must provide intending teachers with every possible opportunity to develop all of the skills that have been described in this book as being required of them in the mixed-ability class and many more. It is also crucial that in-service courses be made available to enable teachers whose initial training did not include such opportunities to be given them now. Mention has already been made of the increased success experienced by those schools whose teachers have been offered this kind of help as part of the advance preparation for the introduction of mixed-ability classes.

However, the most important need is not the acquisition of these skills; it is the development of those powers of judgement that derive from a proper understanding of the enterprise one is engaged in. Such an understanding must be promoted by initial courses of teacher education and it must be fostered by continuing in-service education. In fact, the latter is probably the more important, since it is only when one has become fully involved in the practice of teaching and developed a confidence in one's abilities as a teacher that one can begin to adopt the more reflective approach to education theory that the development of this kind of understanding requires. It is at that point that opportunities need to be provided for teachers to continue their own professional education and, one must add, that all teachers should be required to take advantage of these opportunities.

All curriculum development is teacher development. Without the continuing education of teachers no system, however flexible, will ensure continuous curriculum development of a satisfactory kind. We will only have teachers of the calibre necessary to make this kind of system work when we realize that teachers cannot be trained once and for all time at the beginning of their careers and then left to their own devices, but that they must be helped to develop both their skills and their theoretical understanding of the educational process throughout their careers, so that they can contribute both to the education of each of their pupils and to the continuing development of education itself.

Summary and conclusions

This closing chapter has tried to pick up the issue of flexibility that was made a major feature of that positive rationale for mixed-ability grouping that was offered in Chapter 1 and to reconsider it in the light of the intervening chapters. We noted that this had been the keynote of many of the practical issues that had been discussed and we then looked at some further ways in which it could be used to advantage, considering in particular the allocation of pupils to classes, the problems of dealing with particular subjects and, finally, some of the devices for affecting a smooth initial transition to mixed-ability groupings.

In conclusion, we noted the added responsibility this places on teachers and it was suggested, first, that this is the only basis upon which we can hope to establish a proper system of education in the full sense, and, secondly, that it demands a pattern of teacher education that will, both by initial and in-service courses, produce and maintain teachers of the high calibre such a system requires.

It was suggested in the Foreword that the whole of this book was to be seen not only as offering advice to teachers on the practicalities of teaching mixed-ability classes but also as providing in itself a rationale for that kind of organization. The main feature of that rationale has been sought in the need for a system that is open, loose and flexible enough to allow teachers, through the exercise of their professional judgement, with access to the advice and help of others and in a fully accountable manner, to attend to the educational requirements of every individual child and to the continuing development of education itself. The practice of mixed-ability grouping in schools, where it has been successful, is the nearest we have come to creating such a system and that is its ultimate rationale and justification.

BIBLIOGRAPHY

Allen, J. E. (1971) Pairing and grouping. *Ideas* 21, 10-14.

Associated Examining Board (1971) Pilot project in the assessment of world history at 'O' level. *Ideas* 18, 10-13.

Bantock, G. H. (1968) *Culture, Industrialisation and Education.* London: Routlege & Kegan Paul.

Bantock, G. H. (1971) Towards a theory of popular education. 251-264 in Hooper (1971).

Barker-Lunn, J. C. (1970) *Streaming in the Primary School.* Slough: National Foundation for Educational Research.

Barnes, D., Britton, J. and Rosen, H., editors (1969) *Language, the Learner and the School.* London: Penguin.

Barrs, M., Hedge, A. and Lightfoot, M. (1971) Language in projects. *English in Education* 5, 2, 48-60.

Bernstein, B. B. (1961) Social structure, language and learning. *Educational Research* III, 163-176.

Bernstein, B. B. (1967) Open school, open society? in *New Society* 14 September 1967, also 166-169 in Open University (1971) *School and Society: A Sociological Reader.* London: Routledge & Kegan Paul in association with the Open University Press.

Blenkin, G. M. (1977) The practical validity of an objectives approach to curriculum planning in the infants' school. Unpublished M.A. dissertation, University of London.

Bosworth, D. P. (1971) Programmed science learning. 120-124 in Hardie (1971).

Brown, J. A. C. (1954) *The Social Psychology of Industry*. London: Penguin.

Bruner, J. S. (1960) *The Process of Education*. New York: Random House, Vintage Books.

Bruner, J. S. (1962) *On Knowing: Essays for the Left Hand*. Harvard University Press.

Bruner, J. S. (1966) *Toward a Theory of Instruction*. New York: Norton.

Clayton, B. (1971) Non-streamed secondary mathematics. 108-112 in Hardie (1971).

Connaughton, I. M. (1969) The validity of examinations at 16-plus. *Educational Research* 11, 163-178.

Daniels, J. C. (1961a) The effects of streaming in the primary school I. What teachers believe. *British Journal of Educational Psychology* 31, 67-78.

Daniels, J. C. (1961b) The effects of streaming in the primary school II. Comparison of streamed and unstreamed schools. *British Journal of Educational Psychology* 31, 119-127.

Dearden, R. F. (1968) *The Philosophy of Primary Education*. London: Routledge & Kegan Paul.

Douglas, J. W. B. (1964) *The Home and the School*. London: MacGibbon & Kee.

Downey, M. E. (1977) *Interpersonal Judgements in Education*. London: Harper & Row.

Downey, M. E. and Kelly, A. V. (1975) *Theory and Practice of Education: An Introduction*. London: Harper & Row.

Ellis, J. C. A. E. (1971) Resources. *Ideas* 19/20, 37-40.

Esland, G. M. (1971) Teaching and learning as the organization of knowledge. 70-115 in Young (1971).

Evans, K. M. (1962) *Sociometry and Education*. London: Routledge & Kegan Paul.

Ferri, E. (1971) *Streaming: Two Years Later*. Slough: National Foundation for Educational Research.

Freeman, J. (1969) *Team Teaching in Britain*. London: Ward Lock.

Gannon, T. (1975) Milefield Middle School. 75-92 in Kelly (1975).

Getzels, J. W. and Jackson, P. W. (1962) *Creativity and Intelligence*. New York: Wiley.

Gowan, J. C. et al. (1967) *Creativity: Its Educational Implications*. New York: Wiley.

Haddon, F. A. and Lytton, H. (1968) Teaching approach and the development of divergent thinking abilities in primary schools. *British Journal of Educational Psychology* 38, 171-180.

Hamilton, D. F. (1971) Science for all. 125-131 in Hardie (1971).

Hanson, J. (1971) Management of resources. *Ideas* 19/20, 30-36.

Hardie, M., editor (1971) *At Classroom Level*. Leicester: P.S.W. (Educational) Publications.

Hargreaves, D. H. (1967) *Social Relations in a Secondary School*. London: Routledge & Kegan Paul.

Hargreaves, D. H. (1972) *Interpersonal Relations and Education*. London: Routledge & Kegan Paul.

Haslam, D. J. (1975) Science. 168-189 in Kelly (1975).

Hirst, P. H. (1967) The Curriculum. 75-85 in Schools Council (1967b).

Hirst, P. H. and Peters, R. S. (1970) *The Logic of Education*. London: Routledge & Kegan Paul.

Hooper, R., editor (1971) *The Curriculum: Context, Design and Development*. Edinburgh: Oliver & Boyd in association with the Open University Press.

Horne, M. R. (1975) Fairlop Secondary Girls' School. 40-48 in Kelly (1975).

Hoyle, E. (1969) *The Role of the Teacher*. London: Routledge & Kegan Paul.

Hoyles, E. M. (1975) Vauxhall Manor School. 49-63 in Kelly (1975).

Hunt, A. M. (1975) Sir Leo Schultz High School. 64-74 in Kelly (1975).

Hytch, J. D. and Tidmarsh, D. E. (1971) Mathematics, streamed and non-streamed. A practical approach. 113-119 in Hardie (1971).

Jackson, B. (1964) *Streaming: An Education System in Miniature*. London: Routledge & Kegan Paul.

James, C. M. (1968) *Young Lives at Stake*. London: Collins.

Kaye, B. and Rogers, R. (1968) *Group Work in Secondary Schools*. Oxford University Press.

Keddie, N. (1971) Classroom knowledge. 133-160 in Young (1971).

Kelly, A. V., editor (1975) *Case Studies in Mixed Ability Teaching*. London: Harper & Row.

Kelly, A. V. (1977) *The Curriculum: Theory and Practice*. London: Harper & Row.

Klein, J. (1965) *Samples from English Cultures*. London: Routledge & Kegan Paul.

Kratwohl, D. R. et al. (1964) *Taxonomy of Educational Objectives II: Affective Domain*. London: Longmans.

Lacey, C. (1970) *Hightown Grammar: School as Social System*. Manchester University Press.

Legon, S. K. (1975) Social studies. 140-151 in Kelly (1975).

Liddington, B. (1975) English and drama. 129-139 in Kelly (1975).

Lindesmith, A. and Strauss, A., editors (1969) *Readings in Social Psychology*. New York: Holt, Rinehart & Winston.

Lovell, K. (1967) *Team Teaching*. University of Leeds Press.

Macintosh, H. C. (1970) A constructive role for examining boards in curriculum development. *Journal of Curriculum Studies* 2, 32-39.

Macintosh, H. C. (1971) Towards a strategy of secondary school assessment. *Ideas* 18, 14-17.

Manis, J. and Meltzer, B. (1967) *Symbolic Interaction: A Reader in School Psychology*. Boston: Allyn & Bacon.

Mead, G. H. (1934) *Mind, Self and Society*. University of Chicago Press.

Moreno, J. L. (1934) *Who Shall Survive? Foundations of Sociometry, Group Psychotherapy and Sociodrama.* Reissued (1953) New York: Beacon House.

Musgrove, F. (1966) The social needs and satisfactions of some young people. Part II — At school. *British Journal of Educational Psychology* XXXVI, 137-149.

Musgrove, F. and Taylor, P. H. (1969) *Society and the Teacher's Role.* London: Routlege & Kegan Paul.

Newbold, D. (1977) Ability Grouping — the Banbury Inquiry. Slough: The National Foundation for Education Research.

Northway, M. L. (1968) The Stability of Young Children's Social Relations. *Educational Research* 11, 54-57.

Peters, R. S. (1965) *Ethics and Education.* London: Allen & Unwin.

Pidgeon, D. A. (1970) *Expectation and Pupil Performance.* Slough: National Foundation for Educational Research.

Prettyman, P. A. (1975) Mathematics. 152-157 in Kelly (1975).

Reid, M. R. (1977) Mixed Feelings. *Times Educational Supplement.* 10 June 1977.

Schofield, A. J. (1977) Pastoral care and the curriculum of a comprehensive school. Unpublished M.A. dissertation, University of London.

Schools Council (1965) *Raising the School Leaving Age.* Working Paper No. 2. London: Her Majesty's Stationery Office.

Schools Council (1967a) *Society and the Young School Leaver.* Working Paper No. 11. London: Her Majesty's Stationery Office.

Schools Council (1967b) *The Educational Implications of Social and Economic Change.* Working Paper No. 12. London: Her Majesty's Stationery Office.

Schools Council (1968) *Community Service and the Curriculum.* Schools Council Working Paper No. 17. London: Her Majesty's Stationery Office.

Schools Council (1970) *The Humanities Curriculum Project: An Introduction.* London: Heinemann Educational for the Schools Council.

Schools Council (1971) *A Common System of Examining at 16+.* Examinations Bulletin 23. London: Evans/Methuen Educational for the Schools Council.

Schools Council (1977) *Examination at 18-plus: Resource Implications of an N and F Curriculum and Examination Structure.* Examinations Bulletin 38. London: Evans/Methuen Educational for the Schools Council.

Shaplin, J. T. and Olds, H. F., editors (1964) *Team Teaching.* New York: Harper & Row.

Skilbeck, M. (1976) School-based curriculum development. 90-102 in *Open University Course 203,* Unit 26. Milton Keynes: Open University Press.

Steels, D. L (1969) The Humanities. 112-128 in Kelly (1975).

Stenhouse, L. (1969) The Humanities Curriculum Project. *Journal of Curriculum Studies* 1, 26-33. Also 336-344 in Hooper (1971).

Tansley, A. E. and Gulliford, R. (1960) *The Education of Slow Learning Children.* London: Routledge & Kegan Paul.

Torrance, E. P. (1967) Give the 'Devil' his dues. 137-142 in Gowan (1967).

Vernon, P. E. (1960) *Intelligence and Attainment Tests*. University of London Press.

Vernon, P. E. (1964) *The Certificate of Secondary Education: An Introduction to Objective-type Examinations*. Secondary Schools Examinations Council Examination Bulletin No. 4. London: Her Majesty's Stationery Office.

Walker, B. G. (1975) Art and craft in the middle school. 93-111 in Kelly (1975).

Walmsley, R. S. (1975) French. 190-201 in Kelly (1975).

Warnes, T. (1971) French for everybody. 139-143 in Hardie (1971).

Warwick, D. (1973) *Integrated Studies in the Secondary School*. University of London Press.

White, J. P. (1968) Education in obedience. *New Society*, 2 May 1968.

White, J. P. (1973) *Towards a Compulsory Curriculum*. London: Routledge & Kegan Paul.

White, R. K. and Lippitt, R. (1960) *Autocracy and Democracy: An Experimental Inquiry*. New York: Harper & Row.

Whitehead, A. N. (1932) *The Aims of Education*. London: Williams & Norgate.

Wilson, P. S. (1971) *Interest and Discipline in Education*. London: Routledge & Kegan Paul.

Worthington, F. (1971) The choice of teaching methods for unstreamed classes. 33-44 in Hardie (1971).

Young, A. G. (1975) Northcliffe Community High School. 26-39 in Kelly (1975).

Young, M. F. D., editor (1971) *Knowledge and Control*. London: Collier-Macmillan.

Official enquiries, commissions, reports, etc. quoted or referred to

Board of Education (1926) *The Education of the Adolescent* (The Hadow Report on Secondary Education). London: Her Majesty's Stationery Office.

Board of Education (1931) *Primary Education* (The Hadow Report on Primary Education). London: Her Majesty's Stationery Office.

Board of Education (1938) *Secondary Education with Special Reference to Grammar Schools and Technical High Schools* (The Spens Report). London: Her Majesty's Stationery Office.

Board of Education (1943) *Report of the Committee of the Secondary Schools Examinations Council: Curriculum and Examinations in Secondary Schools* (The Norwood Report). London: Her Majesty's Stationery Office.

Central Advisory Council for Education (1954) *Early Learning*. London: Her Majesty's Stationery Office.

Central Advisory Council for Education (1959) *15 to 18* (The Crowther Report). London: Her Majesty's Stationery Office.

Ministry of Education (1959) *Primary Education*. London: Her Majesty's Stationery Office.

Secondary Schools Examinations Council (1960) *Secondary School Examinations other than the GCE* (The Beloe Report). London: Her Majesty's Stationery Office.

Central Advisory Council for Education (1963) *Half Our Future* (The Newsom Report). London: Her Majesty's Stationery Office.

Department of Education and Science (1964) *Slow Learners at School.* Pamphlet No. 46. London: Her Majesty's Stationery Office.

Inner London Education Authority (1976) *Mixed Ability Grouping: Report of an I.L.E.A. Inspectorate Survey.* London: I.L.E.A.

For many of these documents see **Maclure, J. S.** (1968) *Educational Documents, England and Wales 1816-1967.* London: Chapman & Hall.

INDEX

NOTES